80th SAWE International Conference on Mass Properties Engineering 2021

Virtual Tech Fair

AA005568

Cocoa Beach, Florida, USA
10 - 23 October 2021

ISBN: 978-1-7138-4499-0

Printed from e-media with permission by:

Curran Associates, Inc.
57 Morehouse Lane
Red Hook, NY 12571

Some format issues inherent in the e-media version may also appear in this print version.

Printed by Curran Associates, Inc. (2022)

Society of Allied Weight Engineers, Inc.
Ronald L. Fox
Executive Director
5734 E Lucia Walk
Long Beach, CA 90803-4015
USA

Phone: (562) 596-2873
Fax: (562) 596-2874

ron.fox@sawe.org

Additional copies of this publication are available from:

Curran Associates, Inc.
57 Morehouse Lane
Red Hook, NY 12571 USA
Phone: 845-758-0400
Fax: 845-758-2633
Email: curran@proceedings.com
Web: www.proceedings.com

TABLE OF CONTENTS

SAWE Paper No. 3762
Category No. 12

Artificial Intelligence Techniques for Ship Weight Estimation and Calculation

Upendra Malla
Eastern Shipbuilding Group

For Presentation at the
80th SAWE International Conference on Mass Properties Engineering
October 10-23-2021

Cocoa Beach, Florida

Table of Contents

Tables Listing

Figures Listing

Reference Table

#1: Artificial Intelligence 3rd Edition – Patrick Henry Winston

Abstract:

The advancement of computing and data analysis tools gave rise to the development of Artificial Intelligence (AI) tools. The ship weight estimation and calculation are a simple and tedious process which requires a qualitative weight budgets and quantitative calculations. The evolution of ship weight estimation and calculation using Artificial Intelligence techniques is discussed in the paper and compared with the existing techniques used in the shipping industry. Currently there are several in-house tools and software's which are utilized by design firms and shipyards for the mass properties estimation / calculation, but these tools are not built with any intelligence to make the weight estimate accurate and effective. The implementation of Artificial Intelligence algorithms for the ship weight estimation by considering the constraints like class rules, standards, guidelines etc.

In this paper at the end, it shows the cost and time savings involved by the implementation of Artificial Intelligence techniques in the ship weight estimate program by means of an example AI tool.

History of Ship Weight Estimation:

The ship weight is a critical parameter of the vessel for the concept design, basic design, and construction design phases. To get an accurate estimation of the ship weight right from the early stage of the design is very critical to the functional requirements of the vessel.

Current Methods of Ship Weight Estimation:

In the past and present the ship weight is estimated mainly by the following methods:

1. Hull form co-efficient : The ship weight is estimated by scaling using the ratio of combination of co-efficient of form and fineness.
2. Parent ship data of existing vessels: The historical data of existing vessel is utilized in performing parent ship analysis (regression analysis) to arrive at an estimated ship weight.
3. Deadweight ratio: In the very early stages of the design the deadweight ratio is used to determine the ship weight.
4. Scantling Parameter ratio: The ship weight is calculated using the scantlings of the midship.

It is observed historically in most cases the weight estimation performed utilizing the above methods of estimation is either under or over estimated. To get an accurate weight estimation more parameters / factors need to be applied and thus making the entire weight estimation tedious and complicated.

What is Artificial Intelligence and its techniques:

Artificial intelligence (AI) is the study of the computations that make it possible to perceive, reason and act. It is a science devoted to making machines think and act like humans. Artificial intelligence can be viewed as a blend of engineering and science.

The field of AI research was born at a workshop at Dartmouth College in 1956, where the term "Artificial Intelligence" was coined by John McCarthy to distinguish the field from cybernetics and escape the influence of the cyberneticist Norbert Wiener.

The most widespread and potentially dangerous misconception about artificial intelligence is that it will take away jobs from humans. AI in the workplace has the power to improve how people and businesses perform their jobs, rather than removing the need for humans to do those jobs entirely.

AI Technique is a manner to organize and use the knowledge efficiently in such a way that –

1. It should be perceivable by the people who provide it.
2. It should be easily modifiable to correct errors.
3. It should be useful in many situations though it is incomplete or inaccurate.

AI techniques elevate the speed of execution of the complex program it is equipped with.

Machine learning is a subset of Artificial Intelligence which uses statistical methods to enable machines to improve with experience.

Deep Learning is a subset of Machine Learning composed of algorithms that permit software to train itself to perform tasks. The below figure Fig 1.0 delineates Artificial Intelligence, Machine Learning and Deep Learning.

Figure 1

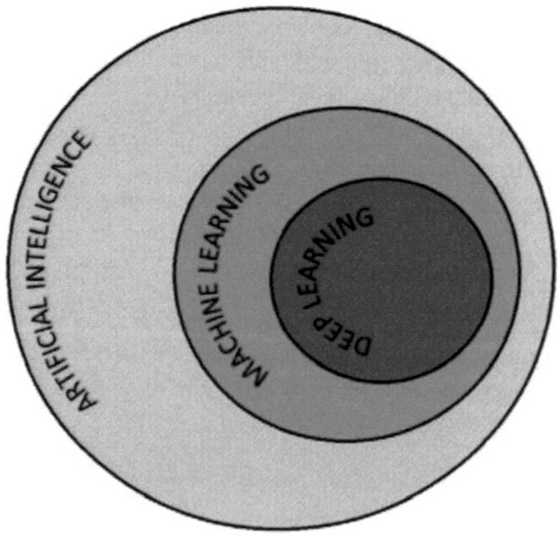

This paper mainly focusses on Machine Learning techniques for the ship weight estimation. The following the different Machine Learning techniques:

1. Supervised Learning: Use supervised learning if you know in advance what you want to teach a machine. This typically requires exposing the algorithm to a huge set of training data, letting the model examine the output, and adjusting the parameters until getting the desired results. You can then test the machine by letting it make predictions for a "validation data set", or in other words, new unseen data.
    a. Classification: Classification algorithms can explain or predict a class value. Classification is an essential component for many AI applications. For example, see Figure 2, classification algorithms can help predict if the equipment crate is open pallet type or fully closed type. The two classes in this case are "yes" and "no". Classification algorithms are not limited to two classes and can be used to classify items into many categories.

Figure 2

OPEN PALLET TYPE
YELLOW PINE WOOD

FULLY CLOSED TYPE
PLYWOOD COMOSITE

Logistic regression is considered the simplest and most basic classification algorithm. A logistic regression algorithm can take more than one input and use the data to estimate the probability of an event occurring. An interesting use of this algorithm can be seen in predicting the type of wood material for the shipping pallet box. The algorithm, in this case, analyzes two different wood material types to estimate the type of wood material probability.

The output is a probable number between zero and one. The number 'one' represents absolute certainty of yellow pine wood, but any number greater than 0.5 predicts the wood material is plywood composite.

    b. Regression: Regression methods are used for training supervised ML. The goal of regression techniques is typically to explain or predict a specific numerical value while using a previous data set. For example, regression methods can take structural weight data for existing vessels, and then predict the ship structural weight for a particular size hull.

Linear regression is considered the simplest and most basic method. In this case, a dataset is modeled using the following equation:

$$y = a\,x + b$$

It is possible to train a regression model with multiple pairs of data, such as x, y. To do this, you need to define a position, as well as the slope of the line, with a minimal distance from all known data points. This is the line that best approximates the observations in the data and can help make predictions for new unseen data. The larger the set of data used to plot the regression analysis, the more accurate will be the AI model.

2. Unsupervised Learning: Unsupervised learning enables a machine to explore a set of data. After the initial exploration, the machine tries to identify hidden patterns that connect different variables. This type of learning can help turn data into groups, based only on statistical properties. Unsupervised learning does not require training on large data sets, and so it is much faster and easier to deploy, compared to supervised learning.
    a. Clustering: Clustering algorithms are unsupervised learning methods. A few common clustering algorithms are K-means, mean-shift, and expectation-maximization. They group data points according to similar or shared characteristics. E.g., Ship weight data plots with steel and aluminum hulls to extract the weight parameter info related to steel or aluminum hull vessel.
    b. Decision Tree: The decision tree algorithm classifies objects by answering "questions" about their attributes located at the nodal points. Depending on the answer, one of the branches is selected, and at the next junction, another question is posed, until the algorithm reaches the tree's "leaf", which indicates the final answer. E.g., Assign a weight factor based on the type of propulsion i.e., diesel engine or electric motor or water jet.

The above AI techniques are utilized in writing the algorithms for the ship weight estimation.

Application of AI in Ship Weight Estimation:

The Artificial Intelligence techniques will be extremely effective with the availability of high computing machines in the present. The ship weight estimation is calculated more accurately by applying the AI techniques for the entire ship weight by following the any standards or guidelines for estimation and reporting. Artificial Intelligence algorithm can be developed with the availability of huge amount of existing vessel data and engineering tools to predict the ship weight.

During the development of the AI algorithm for the ship weight estimation, the criteria or conditions can be added as and when needed to correct the AI model to predict accurate weight estimate for the ship of interest. E.g., For a shipyard for every project the lessons learned, or knowledge gained by the estimation group is lost if the group leaves the company or knowledge from lesson learned not effectively transferred to the next project. With the use of AI tools, the lessons learned, and knowledge is appended in the AI algorithm so that knowledge will be transferred.

Pros and Cons of Artificial Intelligence:

Artificial Intelligence is the ability of a computer program to learn and think. Everything can be considered Artificial intelligence if it involves a program doing something that we would normally think would rely on the intelligence of a human.

The following table, Table 1 shows the pros and cons of using Artificial Intelligence in marine industry.

Table 1
PROS AND CONS OF ARTIFICIAL INTELLIGENCE

PROS	CONS
• Reduces human error	• Not cost effective for small projects with low volume projects
• Optimizes decision making	• Increase machine dependency
• Saves time and costs for large projects	• Drastically changes the work methodology

Artificial Intelligence Examples in Ship Weight Estimation:

There are plenty to implement in the ship weight estimation based on the type of AI technique implemented. The flowchart shown in Figure 3 shows the supervised Artificial Intelligence implementation for the ship weight estimation. If the flowchart avoids the feedback to the model, then the AI model will be un-supervised. This flowchart shows the AI model is developed by writing an algorithm using the existing vessel data and then reducing the error by giving feedback to the model. Once the model is developed fully with all the existing vessel data, then the ship weight estimation is predicted by inputting the vessel of interest data to the AI model.

Figure 3

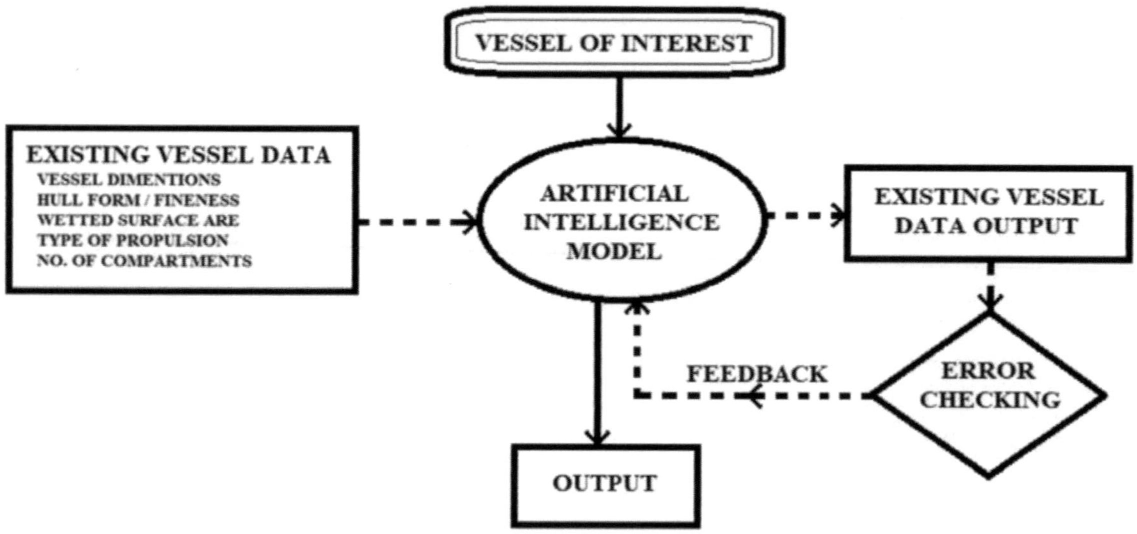

Example 1: Estimation of Steel Structure Weight for a Vessel

Steel weight estimate is based on vessel parameters, no. of compartments, levels, design life etc...is incorporated in the AI algorithm to give an accurate estimate of the weight by saving time.

Example 2: Estimation of tare weight for Equipment

When an item gets delivered to the shipyard, the dimensions, material, and condition of the package is fed to the AI tool, and it will estimate the tare weight. When the item is weighed then the actual tare weight is compared to the estimated tare weight using AI tool and then the difference is fed to the AI which in turn learns. Once the AI learns, the newer estimate from the AI tool becomes more accurate and robust. The advantage with the early estimation of the actual weight using the AI tool will lead to an early detection of anomalous weight increases which would give sufficient time for weight reduction plan.

Conclusions:

The implementation of Artificial Intelligence tools in the marine industry for the estimation of ship weight for various phases of the project will be a cost-effective path for most shipyards and design companies. The AI is currently implemented in most CAD application programs, currently most of the CAD programs have implemented AI to make the modelling more robust and intelligent. The next steps would be the development of commercialization of AI tool for ship weight estimation.

Appendix A

Acronym Listing

AI – Artificial Intelligence
ML – Machine Learning
CAD – Computer Aided Design

SAWE Paper No. 3765
Category Number 31.0

MASS PROPERTIES AND AUTOMOTIVE DIRECTIONAL STABILITY

By

Brian Paul Wiegand, P.E.

(Northrop Grumman Corporation, Retired)

For Presentation at the

80th SAWE International Conference

on Mass Properties Engineering

October 2-6, 2021

Cocoa Beach, Florida, USA

TABLE OF CONTENTS

Chapter:

ABSTRACT

The quantification of automotive directional stability may be expressed through various stability metrics, but perhaps the most basic of these automotive stability metrics is the "**Understeer Gradient**" (**K_{us}**). The Understeer Gradient (in degrees or radians per unit gravity) appears extremely uncomplicated when viewed in its most common formulation:

$$K_{us} = \left[\frac{W_f}{g\,C_{sf}} - \frac{W_r}{g\,C_{sr}}\right]$$

This metric appears to depend only on the front and rear axle weight loads (**W_f, W_r**), and on the front and rear axle cornering stiffnesses (**C_{sf}, C_{sr}**). However, those last quantities vary with lateral acceleration, and the nature of that variation is dependent upon many other parameters of which some of the most basic are: **Total Weight, Sprung Weight, Unsprung Weight, Forward Unsprung Weight, Rear Unsprung Weight, Total Weight LCG, Sprung Weight LCG, Total Weight VCG, Sprung Weight VCG, Track, Front Track, Rear Track, Roll Stiffness, Front Roll Stiffness, Rear Roll Stiffness, Roll Axis Height, Front Roll Center Height, and Rear Roll Center Height**. Note that exactly half of these automotive directional stability parameters as listed herein are **mass properties**.

The purpose of this paper is to explore, through a skidpad simulation, the relative sensitivity of automotive directional stability (as quantified through the Understeer Gradient) to variation in each of the noted vehicle parameters, with special emphasis on the **mass property parameters**.

The simulation is constructed in a spreadsheet format from the relevant basic automotive dynamics equations; the normal and lateral loads on the tires are determined as the lateral acceleration is increased incrementally by a small amount (thereby maintaining a "**quasi-static**" or "**steady-state**" condition). The normal loads are used for the calculation of the lateral traction force potentials at each tire, with the required (centripetal) lateral traction forces apportioned accordingly. From those required (actual) lateral tire forces the corresponding tire cornering stiffnesses are determined; this determination is based upon a tire model developed through a regression analysis of tire test data.

This construction of a fairly comprehensive lateral acceleration simulation from basic automotive dynamic relationships, instead of depending upon commercial automotive software such as "**CarSim**" (vehicle model) and **Pacjeka** "**Magic Formula**" (tire model), constitutes a unique aspect of this paper; this return to basics hopefully provides a clearer view and understanding of the results than would be the case otherwise. Even more unique is this paper's emphasis on, and exploration of, the role specific **mass property parameters** play in determining automotive directional stability.

1 – HISTORY OF AUTOMOTIVE DIRECTIONAL STABILITY THEORY DEVELOPMENT

Control of vehicle motion requires understanding of the directional stability in the principle directions of motion. The principle directions of vehicle motion are conventionally taken as being along the vehicle longitudinal, lateral, and vertical axes; consequently directional stability involves the rotational motion about those axes, resulting in what is known as the roll, pitch, and yaw stability, respectively. For aircraft the principle directional stability concern is pitch stability[1], but for automobiles the principle directional stability concern is yaw stability. Although the development of the automobile preceded the aircraft, the development of automotive directional (yaw) stability theory lagged the development of aircraft directional (pitch) stability theory by about 35 years[2]; when the study of automotive stability was finally undertaken in a scientific way that study tended to draw heavily upon the aeronautical stability theory already established[3].

The primary forces which determine the dynamic behavior of aircraft are aerodynamic forces generated by pressure differentials acting over the aerosurface areas. In contrast, the primary forces which determine the dynamic behavior of automobiles are the friction (more appropriately called "traction") forces acting over the tire-to-road contact areas; it is these forces that accelerate, decelerate, and maneuver the automotive road vehicle. Tire behavior was initially best known for attenuating the shock and vibration effects of road surface irregularities on the vehicle, its cargo, and passengers. The absolutely critical role that tires play in providing vehicle directional stability was not understood until much later. The complex pneumatic-elastic rubber and cord structure known as the tire underwent innumerable improvements and refinements for many years without being fully understood in its mechanisms and behavior.

Robert William Thomson patented (**UK**) the first pneumatic rubber tire in 1845. Thomson carried out crude experiments demonstrating the benefits of using pneumatic rubber tires, instead of the then conventional rigid wheels, on horse drawn conveyances. Not only was

---

[1] Early on in the development of aircraft it was noted that birds have their wings mounted above the center of gravity (**CG**) of their bodies; almost immediately aircraft were designed with the aerodynamic center of pressure (**CP**) above the **CG**, either by high-mounting the wings and/or dihedral. Thus roll stability was obtained, and as for yaw stability the provision of an adequate tail fin and/or rudder was all that was needed.

[2] Reference [19], pg.123. Actually, the study of aircraft directional stability long preceded the existence of any aircraft; Sir George Cayley (1773–1857) published **On Aerial Navigation** in three parts over 1809-1810. Frederick William Lanchester (1868-1946) published **Aerial Flight** in two parts over 1906-08, and George Hartley Bryan (1864–1928) published **Stability in Aviation** in 1911. All of these studied the principles of flight long before powered aircraft became a reality. The Wright's were well aware of the works of Cayley, and possibly of Bryan, but initially they still flew a highly unstable "**Wright Flyer**" (Kitty Hawk, 1903).

[3] In 1953 A.G. Fonda of the Cornell Aeronautical Laboratory (**CAL**) explicitly used aircraft stability concepts for creation of a linear automotive stability model. In1955 Leonard Segel, William F. Milliken, David W. Whitcomb, et al, of the Cornell Aeronautical Laboratory greatly advanced the theory, culminating in the seminal 1956 automotive stability paper attributed to Segel: "**Theoretical Prediction and Experimental Substantiation of the Response of the Automobile to Steering Control**", which was presented to the Institution of Mechanical Engineers in London. on 30 August 1956.

the ride quality improved, but the drawbar pull necessary to maintain motion was decreased[4]. However, this indication regarding the relative rolling resistance of tires with respect to that of rigid wheels would represent the extent of tire knowledge for a long time, and the prevailing level of technology of the period would prevent the general adoption of Thomson's invention.

When John Boyd Dunlop re-invented the pneumatic tire (1888, **UK** patent) the prevailing level of technology had risen considerably and the bicycle craze was in full swing, with the rise of the automobile just beginning. For approximately the next forty years the pneumatic rubber tire would undergo a rapid development unaided by any scientific understanding of the physical mechanisms involved. Such an understanding would await the development of machines which would allow for the testing of tire behavior in a laboratory environment[5]. A short history of the drive to a scientific understanding of the physical mechanisms of tires, and thereby to an understanding of the automobile, may be outlined as follows.

It was by 1925 that the French researcher George Broulheit was able to identify the tire lateral behavior characteristic of "**slip angle**" in his paper on front axle "**shimmy and tramp**" ("**La Suspension de la Direction de la Voiture Automobile: Shimmy et Danadinement**")[6], which was the automotive problem of the day[7]. The relationship of tire "**slip angle**" to lateral traction forces was to prove essential to all future understanding of automotive lateral behavior[8].

Just prior 1931 what was possibly the first tire testing machine, a rotating steel drum tire-tester, was built by the German researchers Becker, Fromm, and Maruhn[9]. They too were investigating the problem of "**shimmy and tramp**" as a steering matter ("**Schwingungen in Automobillenkungen**"), and had realized that if they were to come to any exact understanding of lateral automotive dynamic behavior then they would need solid quantitative tire data.

By 1934 Maurice Olley of General Motors developed the automotive behavior concepts of "**over**", "**under**", and "**neutral**" **steering**, thereby providing a qualitative framework for characterizing automotive yaw behavior ("**Stable and Unstable Steering**"), and advanced the use of a circular track called a "**skidpad**" for the empirical evaluation of vehicles with regard to those behavior characteristics. He also advanced the concept of a "**critical speed**" beyond which an initially mildly "**oversteering**" vehicle becomes uncontrollable[10].

---

[4] Thomson published an article on the relatively low resistance to motion of "***tyred***" wheels with respect to rigid wheels in **Mechanics Magazine** (Glasgow) sometime in 1846; no copies appear to be extant today.

[5] Reference [29], pg. 6.

[6] Reference [20], pg. 70. Reference [29], pg. 7. Reference [27], pg.218.

[7] The prior automotive problem(s) of the day had to do with the nature of the major ride motions and possible interaction between them; researchers H.S. Rowell (1922) and J.J. Guest (1925) had independently identified the fundamental causes of poor ride (slow "***beats***" leading to "***car sickness***", sudden lurches and jumps) experienced in certain vehicles of the period.

[8] Reference [29], pg.7. Reference [27], pg 218. Reference [20], pg. 70.

[9] Reference [20], pg. 70. In the **UK** one of the first wind tunnels was built at the National Physical Laboratory (**NPL**) in 1903, and the Wright Brothers built a small simple one of their own even earlier.

[10] Reference [20], pp. 559-560. Reference [34], pp. 372-373. Reference [29], pg. 7.

Shortly afterward in 1935 R.D. "Cap" Evans of the Goodyear Tire & Rubber Company published the results of quantitative investigation of tire behavior via another rotating drum (84 inch diameter) tire-tester[11]. His tire data conclusions ("**Properties of Tires Affecting Riding, Steering, and Handling**") were to prove invaluable for later researchers developing mathematical models of tire behavior.

During the years 1939-1945 the advent of World War II stopped most research into tire and automotive behavior, but B. Von Schlippe and R. Dietrich at the Junkers aircraft company developed ("**Das Flattern des Pneumatischen Rades**", 1941) a simple tire model ("**taunt massless string on elastic foundation**")[12].

However, around 1939 Dr. A.W. Bull at **US** Rubber would publish tire data ("**Tire Behavior in Steering**"), and around 1954 S.A. Lippman, also at **US** Rubber, would also develop some tire model insights[13]. The significance of these tire models was that it could be shown at highway speeds and small slip angles the **transient** phase of tire adjustment to lateral force could be neglected so that the results of modeling the **steady-state** phase of automotive behavior would have broad validity; while this did not mean that interest in the **transient** phase would be abandoned it did greatly influence the direction of further analysis and modeling[14].

In 1950 British automotive engineer G.E. Lind Walker introduced the concepts of "**neutral steer point**" and "**static margin**" into the discussion of automotive directional stability ("**Directional Stability, a Study of Factors Involved in Private Car Design**"), adapting these concepts from existing aircraft stability theory[15].

By 1953 Robert Schilling of General Motors, in conjunction with/based on the work of colleagues Maurice Olley, Ken Stonex, et al, derived a set of basic differential equations for automotive motion. Through a visit by Cornell Aeronautical Laboratory (**CAL**)[16] researcher William Milliken, the **GM** research personnel became aware of the **CAL** development work on the equations of motion for aircraft, resulting in a **GM** contracture of **CAL** for further development and refinement of the automotive equations; **CAL's** Albert G. Fonda then used aircraft stability concepts to produce a linearized automotive model. This redirection of aircraft dynamics experts toward automotive matters would prove to have profound consequences.

---

[11] Reference [20], pg. 70. Evans conducted these famous "first in the **US**" tire force and moment tests during 1932-1933 at the urging of Maurice Olley.
[12] Reference [19], pg. 424.
[13] Ibid, pp. 420-421.
[14] Ibid, pg. 424.
[15] Ibid, pg.127.
[16] The Cornell Aeronautical Laboratory got started as private defense contractor Curtiss-Wright's aero laboratory in WW II. Post-war, Curtiss-Wright donated the lab to Cornell University to operate "***as a public trust***". The lab operated under the name Cornell Aeronautical Laboratory from 1946 until 1972. During the late 1960's/early 1970's universities came under criticism for conducting war-related research, and consequently in 1972 Cornell reorganized the lab as the "***for-profit***" Calspan Corporation and sold all its stock to the public.

In 1954 Prof. Yves Rocard of the Sorbonne published an analysis (**"L'instabilité en Mécanique; Automobiles, Avions, Ponts Suspendus"**) of automotive stability based on a four wheel rigid model (no roll effects) without steering or normal load variation at the wheels[17]; this may be regarded as the continuation of the line of French research into automotive directional stability that began with Broulheit. One of the key results of Rocard's analysis was the indication of the possible existence of a "**critical speed**". It was stated that "**for speeds up to this value the motion will be stable but for higher speeds it will be unstable**"[18].

Soon afterward in 1955 French researchers M.A. Julien and G.J. Arnet produced an analysis of a less simplified model that was separated into a **sprung and an unsprung mass**; this was a major step closer to reality, but today the **unsprung masses** for conventional automotive configurations are generally treated as four independent **masses**. A key result of this analysis was the indication "**that there may be several critical speeds**"[19].

By 1956 Leonard Segel, William F. Milliken, David W. Whitcomb, et al, of the Cornell Aeronautical Laboratory would greatly advance understanding of tire behavior and automotive directional stability. Tire data was independently generated due to the construction of the **AF-CAL** on-road tire-tester; this device was built under contract to the **US** Air Force (hence "**AF**") to initially study the behavior of aircraft tires. The **CAL** team's efforts would culminate in the seminal 1956 paper attributed to Leonard Segel: "**Theoretical Prediction and Experimental Substantiation of the Response of the Automobile to Steering Control**"[20].

The model in this 1956 paper had four wheels on two axles (yet seems to have some of the simplifications of a "**bicycle model**") with the vehicle again divided into a **sprung and an unsprung mass**; some of the effects of roll (held to "quite small" values about a fixed roll axis parallel to the ground plane) and load transfer were thereby taken into account, but steady-state conditions prevailed and tire slip angle and other functions were linearized (with normal loads held constant!) such that the stability analysis results were valid only up to about 0.3 g in lateral acceleration. However, this paper represented the culmination of much that had gone before, and marked the beginning of modern automotive stability theory[21]. The theory as developed in this paper was validated experimentally by use of an instrumented 1953 Buick Super.

Additionally in 1956 David W. Whitcomb wrote a paper that also (ala Walker) adapted the aircraft concept of "**static margin**" to the automobile, and explored automobile stability and control using a two degree of freedom (**DOF**) model including yaw and side-slip which is

---

[17] Reference [27], pp. 237-243.
[18] Ibid, pg. 242.
[19] Ibid, pg. 243.
[20] Incidentally, Segel's paper was presented under the auspices of the **IME** in London instead of the **SAE** in Detroit because the **SAE** had turned down the paper for presentation due to its being "***too mathematical and of academic value only***"! Reference [37], pg.542.
[21] Reference [27], pp. 243-261. Taborek states somewhere in his book (Reference [28]) that systematic study of automotive directional stability had begun <u>only</u> in the "***last decade***" (1950's).

referred to as the "**bicycle model**". This simple model, the origin of which possibly goes back to the 1930's, continues to prove very useful in development of automotive dynamic concepts.

In 1957 Leonard Segel published another paper ("**Research in the Fundamentals of Automotive Control and Stability**") which presented a review of the state of the art, described the development of the lateral equations of motion and subsequent empirical verification, and concluded with explanation of certain basic steady-state and dynamic properties.

After the 1950's the rise of ubiquitous digital computer resources ultimately resulted in the modern day plethora of highly detailed and complex models. Initially programming was done on "mainframes", the use and capability of which was very limited. Much computing was done on analog instead of digital computers; although the analog computers tended to be more limited in application they were also much faster. To combine the best features of analog and digital computers a class of computers known as hybrid computers arose which were quite popular for a while, but as advances in electronics decreased the size and expense of digital computers, while simultaneously increasing the power and speed, the digital computer became dominant[22]. There was also increased accessibility as batch processing gave way to time-sharing, and time-share gave way to personal computing. The intricacies of arcane system languages like **JCL** (Job Control Language) ultimately gave way to icon/menu oriented, "mouse" driven systems. First generation binary or hexadecimal programming languages gave way to second generation (like **Assembler**), and then third generation (**FORTRAN**, **BASIC**, etc.) programming languages. The end product of all this electronic evolution has been to place within the reach of the general computer-using population a power that was not there before[23], reducing the need for many of the simplifications previously required in automotive dynamic modeling, along with much else[24].

In 1960 Martin Goland and Frederick Jindra of the Southwest Research Institute added the roll **DOF** to the two **DOF** (yaw and sideslip) vehicle model of Segel in order to more realistically determine the normal loads on the tires ("**Directional Stability and Control of a Four-Wheeled Vehicle in a Flat Turn**"). With the addition of roll the effects of load transfer and the variation of the cornering stiffness of the tires with normal load could be more realistically accounted for. By means of their simulation Goland and Jindra explored the effect on stability of varying the **longitudinal center of gravity** (**LCG**)[25], tires, and inflation pressures. They were able to show that the directional stability of the vehicle would vary depending on speed ("**V**") and radius ("**R**") of the turn[26].

[22] Reference [6], pg. 38.

[23] Reference [30], pg. 1.

[24] Students involved in the **SAE** Formula Student Design Competition (**FSAE**) commonly utilize sophisticated software such as **CATIA**, **ANSYS**, **LS-DYNA**, **Auto-CAD**, **SolidWorks**, **MatLab**, **Simulink**, **STAR-CCM+**, **Pro/ENGINEER**, **CarSim**, **SuspensionSim**, **ADAMS/Car**, etc.

[25] However, they confuse the effects of individual parameters by varying the front-to-rear roll stiffness balance in sync with the varying of front-to-rear **weight balance**; they should have varied each parameter individually.

[26] Of course, the lateral acceleration "$\mathbf{a_y}$" is determined by those parameters: "$\mathbf{a_y = V^2/R}$".

In 1965 Ford Motor Company's Walter Bergman published a paper ("**The Basic Nature of Vehicle Understeer-Oversteer**") that departed from the then conventional practice by defining **understeer-oversteer** in the **transient** as well as in the **steady-state** condition; "***...he noted that understeer and oversteer could be recognized by considering the change in the yaw velocity induced by a change in lateral acceleration***.[27]" In an examination of all conceivable automotive stability case possibilities Bergman also discussed **static** versus **dynamic** (termed "**oscillatory**") stability, noting that "***...a commercially acceptable understeering vehicle is an oscillatorily stable vehicle. However, an understeering vehicle might become oscillatorily unstable due to...unfavorable...operating conditions. An oversteering vehicle does not develop an oscillatory response...***"[28]. Due to its nature, an "**oscillatory**" response belongs to the transient realm, and most drivers who have experienced such a response will testify that the oscillation will generally die out (decay) in short order if matters are not complicated by panicked actions at the steering wheel.

In 1967 R. Thomas Bundorf of General Motors published the relation of certain vehicle design parameters to the "**characteristic speed**" and to "**understeer**" ("**The Influence of Vehicle Design Parameters on Characteristic Speed and Understeer**"). Bundorf argued that under most normal driving conditions, which he characterized as having lateral accelerations below one-third a standard gravity value, a vehicle can be accurately modeled by a linear model.

In 1973 independent researcher and inventor Frank H. Speckhart presented a paper based on a **computer simulation** utilizing a vehicle model containing fourteen **DOF** ("**A Computer Simulation for Three-Dimensional Vehicle Dynamics**")[29]. Six **DOF** were assigned to the **sprung mass**, four **DOF** were assigned to the suspension movement at the four corners of the vehicle, and four rotational **DOF** of freedom were assigned to the wheels; with a model having that many **DOF** it was fortunate that Speckhart had adequate digital computer resources available. Furthermore, the tire model utilized nonlinear characteristics, and the suspension model included camber and roll stiffness with an inclined roll axis. In the paper simulated results of automotive "**J-turn**" maneuvers were compared with empirical results for two different speed and steering inputs.

---

[27] Reference [29], pg. 9. While this is an early recognition of stability in the **transient** phase, the industry emphasis has continued to be mainly on stability within the **steady-state** phase.

[28] Reference [3], pg. 4. The statement that dynamically unstable vehicles do not reach the public (are not "***commercially acceptable***") is reassuring but not enlightening. As to how a "***commercially acceptable understeering***" vehicle might become dynamically ("***oscillatorily***") unstable due to "***operating conditions***", consider such a vehicle compromised by an ignorant overloading which shifts the **CG** far aft and perhaps upward as well, then driven at a high rate of speed into a tight turn....Of course, manufacturers should make allowances for a certain amount of such ignorant useage in the design of their product, and generally do to some extent.

[29] Reference [29], pg. 9. By 1976 Speckhart was marketing a general purpose program ("**CSMP**") for "**continuous system modeling**".

In 1987 Dr. Andrzej G. Nalecz of the **NHTSA** tested, by **microcomputer simulation**, twenty-five different suspension types to understand the effects of suspension design on the stability of light vehicles ("**Investigation into the Effects of Suspension Design on Stability of Light Vehicles**"). A typical three **DOF** lateral dynamics model was used with the addition of a pitch degree of freedom. The roll axis of **sprung mass** was not fixed but instead varied as a function of body roll.[30]. This set of simulated tests enabled Nalecz to argue that the assumption of a fixed roll axis could not be justified for certain types of suspensions such as double wishbone and MacPherson strut.

In 1992 Dr. Andrzej G. Nalecz utilized a more complex model with eight **DOF**, consisting of a three **DOF** lateral dynamics model coupled to a five **DOF** planar rollover model, to developed a more comprehensive understanding of lateral behavior ("**Development and Validation of Light Vehicle Dynamics Simulation** [**LVDS**]")[31]. The lateral dynamics model was derived in the same manner as Segel's original model; the roll model consisted of **sprung and unsprung masses** connected through the various elements of the suspension system. However, now the effects of aerodynamic forces and moments, plus the full effects of lateral and longitudinal "**weight transfer**"[32], were also included in the model.

Today automotive simulations of many degrees of freedom and great complexity are routinely run for all sorts of scenarios: acceleration, braking, cornering, and any combination thereof. Simplifications such as: fixed roll axes, taking camber effects as negligible, ignoring tire aligning moments, modeling as only a two **mass** system (**sprung** and **unsprung**), ignoring aerodynamic forces and moments, linearizing tire functions, ignoring suspension compliances, etc., etc. are now largely unnecessary. Very often a mathematical model of the driver, itself a matter of great complexity, may be included in an overall automotive system model. Dynamic simulations are now employed to not only investigate automotive behavior, but to control it via onboard processors in the form of **ABS** (Anti-skid Braking Systems), **SCS** (suspension control systems), **TCS** (traction control systems), and **ESC** (electronic stability control) systems.

Now that personal computers are so widely available, and the theory of automotive directional stability is so well developed, any moderately intelligent individual can construct an automotive simulation on a par with, or even better than, those that the great minds of automotive dynamics created during the historical course of theory development. Creation of a simulation such as this paper is based upon can be remarkably reflective of the physical reality,

---

[30] Reference [29], pg. 10. Kinematic analysis of the suspension systems with roll can produce a model of how the roll centers, and in turn how the roll axis, will vary with roll. Presumably this is what Nalecz did.

[31] Ibid. Nalecz's **LVDS** program would soon join the growing ranks of commercially available software for automotive analysis. **LVDS** was capable of predicting light vehicle maneuvers from low lateral acceleration cornering to spin out and rollover.

[32] "**Weight transfer**" is somewhat of a misnomer, as the situation is really more of a "**load transfer**"; no vehicle **mass** is relocating internally (except for "**free surface**" effects, etc.).

but this author must share the credit for this creation with the many that struggled long before when the going was not so easy.

2 – AUTOMOTIVE DIRECTIONAL STABILITY THEORY

As noted earlier, automotive stability begins with the tire. Application of a lateral force on a tire not only causes a distortion of the tire carcass diminishing the tire/road area contact, but there are other effects as well. When a tire moving with velocity "**V_o**" encounters a side load there is a consequent sideways motion. The resultant new net motion "**V**" is the combination of the original motion "**V_o**" and the motion resulting from the side load. The angle "**ψ**" between this new direction "**V**" and the original direction "**V_o**" is called the "**slip angle**". This term is a misnomer as it gives an erroneous impression[33]; the tire is not necessarily slipping or sliding in the direction of the side load.

What is actually happening is that there are a series of small lateral movements "**d_y**" of the tire due to the cyclical distortion of portions of the carcass as those portions come into contact with the road as the tire rolls forward. The combination of the original forward rolling velocity and all those infinitesimal "**side steps**" results in the new velocity direction "**V**". This is the situation illustrated in Fig. 2.1. This scenario of a change in direction due to a side load, possibly a wind gust or an inertial load due to an uneven road, may be called the "**directional disturbance scenario**". The occurrence of a disturbance force causing tire carcass distortion, "**slip angle**", and consequent change in direction, represents only one particular order of "**cause and effect**".

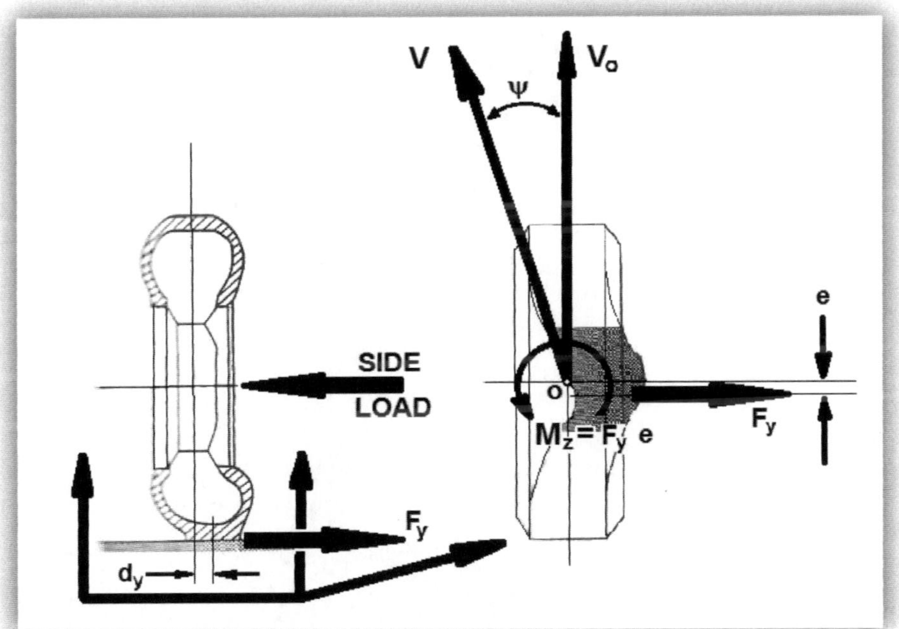

Figure 2.1 – DIRECTIONAL DISTURBANCE PRODUCES "d_y", "V", "ψ", "F_y", & "M_z"

[33] Reference [12], pg. 348. Reference [2], pg. 73. In this last reference famed automotive engineer and author Donald Bastow suggests that a better term for "ψ" might be "**drift angle**", an idea this author has enthusiastically endorsed but admittedly only sporadically adhered to.

There is another "**cause and effect**" scenario wherein a "**slip angle**" is intentionally created resulting in a traction force inflicting a turning motion on the vehicle; this scenario is called the "**turning scenario**". Unlike the random happenstance of directional disturbance, the "**turning scenario**" begins with the intentional act of a driver cranking over the steering wheel, thereby rotating the front tires through the steering angle "**δ**". As the tires rotate through the angle "**δ**" the portion of the tire in contact with the ground becomes distorted giving rise to a lateral shear force distribution of which the resultant centripetal force "**F**" tends to drive the tire towards the turn center "**TC**" of a turn of radius "**R**".

Fig. 2.2 is an illustration of this turning scenario; note that the lateral sheer stress distribution is such that the resultant lateral traction force "$\mathbf{F_y}$" acts at a distance "**e**", called the "**pneumatic trail**", aft of the tire center "**o**" (which was also true in the previous scenario). This offset of the lateral force from the tire center results in a moment "$\mathbf{M_z}$", called the "**self-aligning torque**", about that center. The generation of the angle "ψ" and the moment "$\mathbf{M_z}$" are two of the most significant aspects of tire lateral behavior. The combination of the original direction "$\mathbf{V_o}$" plus some new motion toward the turn center "**TC**" produces a new resultant direction "**V**" operating at some "**slip angle**" "ψ" to the longitudinal axis (apparent direction) of the tire. The operation of the centripetal force "**F**" generates an equal and opposite centrifugal virtual "**side load**" force in accord with the concept of dynamic equilibrium; this side load force is proportional to the **mass** supported by the tire times the radial acceleration: "$\mathbf{m\ V^2/R}$".

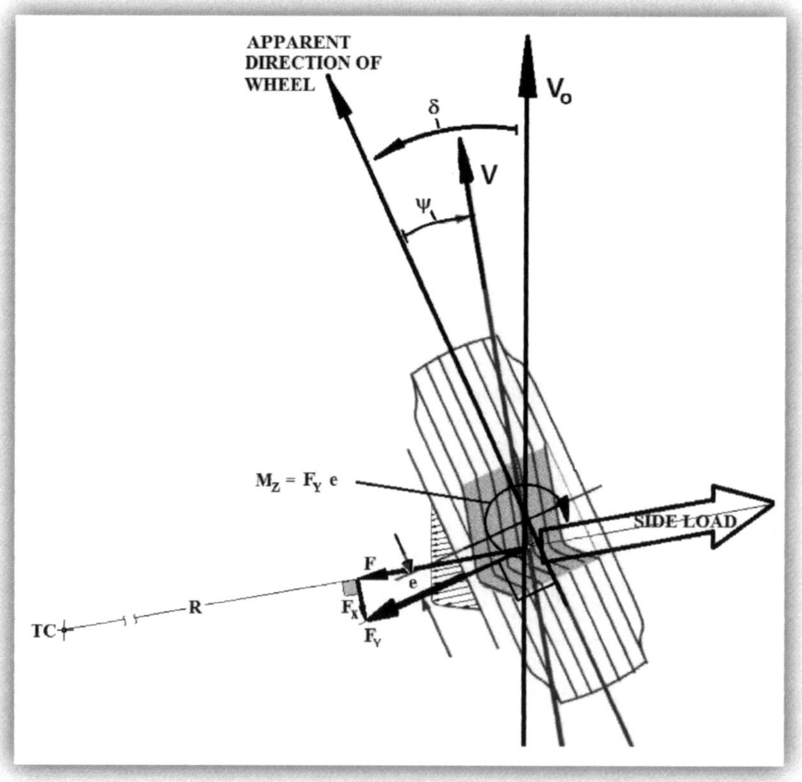

Figure 2.2 – STEERING INPUT PRODUCES "δ", "ψ", "V", "F_x", "F_y", & "M_z"

Because "**F**" must lie along the turn radius (just as "**V**" must be orthogonal to the radius) it is not truly lateral in orientation with respect to the tire; it may be resolved into two components: "$\mathbf{F_x}$" and "$\mathbf{F_y}$". The "$\mathbf{F_y}$" is the true lateral traction force, while "$\mathbf{F_x}$" may be referred to as the "**turning drag force**". The existence of this drag force is why, even without any braking or easing up on the throttle, there will be some loss of velocity for a vehicle undergoing a turning action. Lastly it should be noted that the action of the "**self-aligning torque**" "$\mathbf{M_z}$" is such that in this case the tendency is to return the steered wheel back to its original direction "$\mathbf{V_o}$"; this "**self-aligning torque**" is an important factor in vehicle steering and control[34].

With regard to these two scenarios of lateral tire behavior the observation has often been made as to how it is difficult to discern between cause and effect: does the lateral force cause the formation of a slip angle, or does the formation of a slip angle lead to the creation of a lateral force? It is hoped that the previous careful expositions regarding the two possible scenarios will clarify things, but it doesn't really matter[35]. Regardless of whether the force creates the acceleration, or the acceleration creates the force, it is sufficient to know the relation "**F = m a**"; the rest is philosophy.

However, it is of importance in constructing computer simulations of dynamic effects to not only have a clear definition of cause and effect, but to have some estimation of any time lag between the two: the time lag between when the driver cranks the front tires over to the steering angle "$\boldsymbol{\delta}$" and when the tires settle into the steady state values of "$\boldsymbol{\psi}$", "**V**", "**F**", and "$\mathbf{M_z}$" is generally equal to the time it takes for the tire to undergo a rotation of about 180 (π radians) to 360 degrees (2π radians)[36]. This lag time can be important in a simulation; remember that it was the theoretical model indication of tire response times being insignificant that triggered an emphasis on investigation of steady-state automotive behavior in the 1950's.

Since the slip-angle/lateral-force relationship is dependent upon quite a few parameters (inflation pressure, normal load, temperature, velocity, etc.), it is common to look at functions which constitute only a partial differential of the total relationship (for which no one has yet established a complete definitive formulation based on physics[37]) in order to achieve a degree of

---

[34] Reference [12], pp. 359-361. The "**self-aligning torque**" varies with "$\boldsymbol{\psi}$" (proportional to "$\mathbf{F_y}$") and "**N**", and therefore typically has units of "***...approximately 0.033 ft-lb/lb/deg for bias tires, and 0.043 ft-lb/lb/deg for radial tires...***". This torque typically increases "***up to about 8 degrees of slip angle. At greater angles...decrease in aligning moment...At very high slip angles...can actually become negative.***"

[35] Reference [19], pg. 19: "***...lateral...force may be thought of as the result of slip angle, or the slip angle as the result of lateral force...***".

[36] Reference [12], pg. 349. During the tire rotation required (usually 63%) to reach a steady-state or near steady-state condition a certain distance will be traversed by the automobile; this distance is referred to as the "**relaxation length**", and the time taken to cover that distance is the "**relaxation time**".

[37] Hans Bastiaan Pacjeka, Professor Emeritus at Delft University of Technology in the Netherlands, has developed a tire model called the "**Magic Formula**" because it is based on relatively little underlying physics; instead it is mainly based on a regression analysis of reams of empirical tire data.

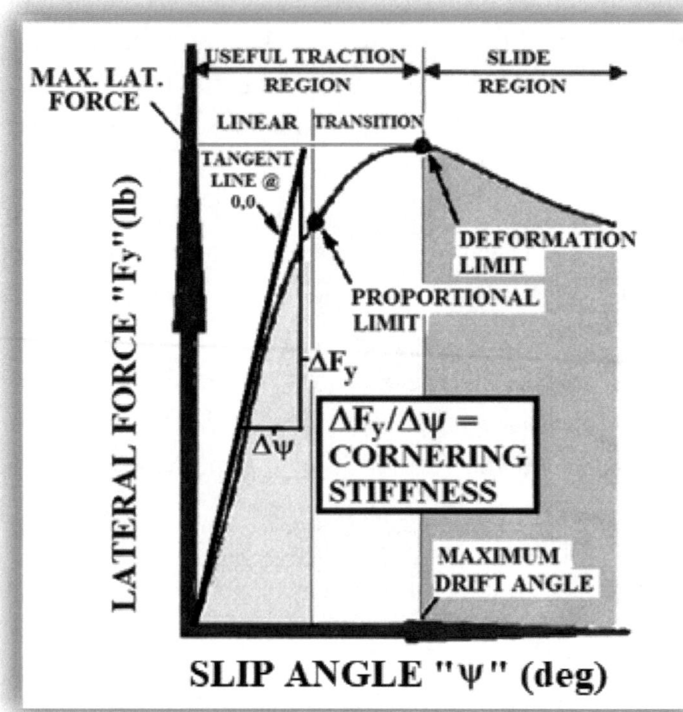

Figure 2.3 – TIRE LATERAL FORCE vs. SLIP ANGLE @ CONSTANT "N"

understanding. If the tire slip-angle/lateral-force partial differential function (with inflation pressure "$\mathbf{P_i}$" and normal load "**N**" held constant) is plotted the result looks like Figure 2.3[38]:

The lateral force increases with slip angle (or vice versa) on a shallow curve, reminiscent of some material stress-strain curves, up to a deformation limit. At that point the situation is very unstable and will either quickly return to the safety of the "useful traction region" or result in 100% slip at the tire/road contact patch (a "**slide**" or "**skid**").

If a portion of the function is linearized by the fitting of a straight line to the original shallow curve, then a simplified lateral-force/slip-angle relationship (the slope of the fitted line) can be obtained. This quantity is commonly called the "**Cornering Stiffness**" and is symbolized as "$\mathbf{C_s}$" (sometimes it is also called the "**Cornering Power**", but that is yet another misnomer like "**slip angle**" or "**shock absorber**"). Linearized tire relationship values such as "$\mathbf{C_s}$" are often used in directional stability determinations and for various simulations and studies of automotive dynamic behavior.

Note the tangent line at point "**0,0**"; the equation for this straight line was often used by classical automotive dynamicists to serve as the linearization of the lateral force-slip angle function[39]. Since this linearization is not a good representation at "high" slip angles the

[38] Reference [12], pg. 350.
[39] Reference [12], pp. 198-200. On page 350 this reference states that "***…C_s is the…slope of the curve at zero slip angle…***" (even though that would not seem to be a very good fit to the curve). Although cars turn left and right,

theoretical results obtained through such linearized modeling were good only up to about 0.3 g's in lateral acceleration. Due to the ubiquity of present day computer resources such gross simplifications are no longer necessary; if the function "**ψ=f(Fy) @ N**" is known then the slope can be found at a every point along the **ψ=f(Fy)** function. The totality of tire behavior can be modeled as consisting of a number of such functions, covering an anticipated working range of normal load values:

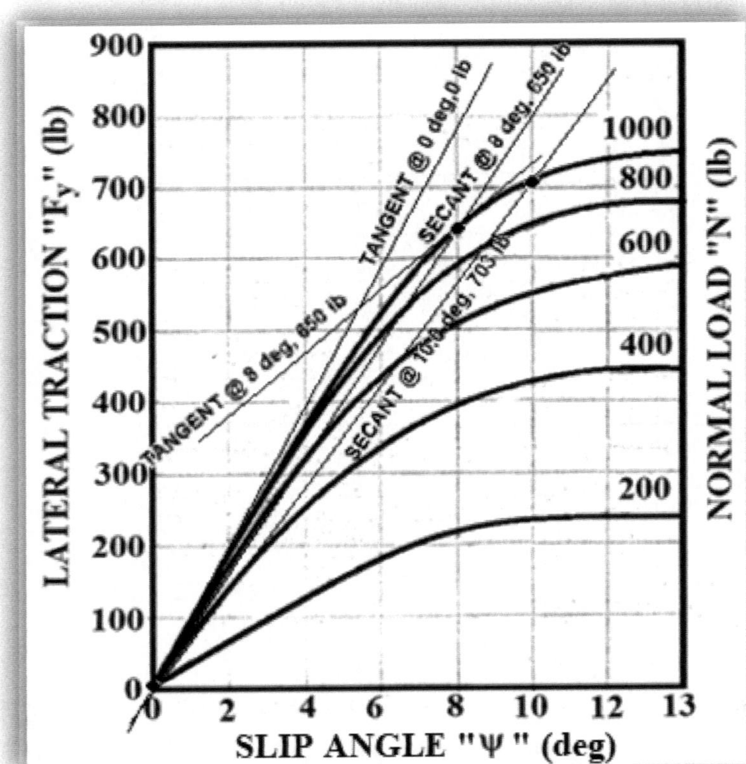

Figure 2.4 – TIRE LATERAL FORCE vs. SLIP ANGLE @ CONSTANT "N" LINEARIZATION

The Fig. 2.4 illustrates some tangent lines to the traction force–slip angle functions at certain constant normal loads. For the **N=1000** function the tangent at the "**0,0**" point is shown, which, as noted, in the past may have been taken as the linear representation of that function. As a possible alternative linearization for the **N=1000** function the secant line from "**0,0**" point to the "**8,650**" point might have been used, or maybe the secant line from "**0,0**" point to the "**10,703**" point. Obviously, the way the linearization is obtained will have a big impact on the accuracy of the representation, but nothing can beat having an actual "**ψ=f(Fy)**" equation so that the tangent at each and every point along the function may be found by evaluating the partial differential "$\partial F_y/\partial \psi$" at that point.

implying positive and negative "C_s" values, by **SAE** convention "C_s" is always positive. "C_s" has units such as "**lb/deg**", but when normalized by the normal load "**N**" the "**Cornering Stiffness**" becomes the "**Cornering Coefficient**" "C_c" in units of "**lb/deg/lb**".

With a chart of empirically obtained data for a specific tire, if the actual lateral forces at the tire are known, then the slip angles in response to those lateral forces can be determined. Such a source of specific tire information may look as per the following figure[40]:

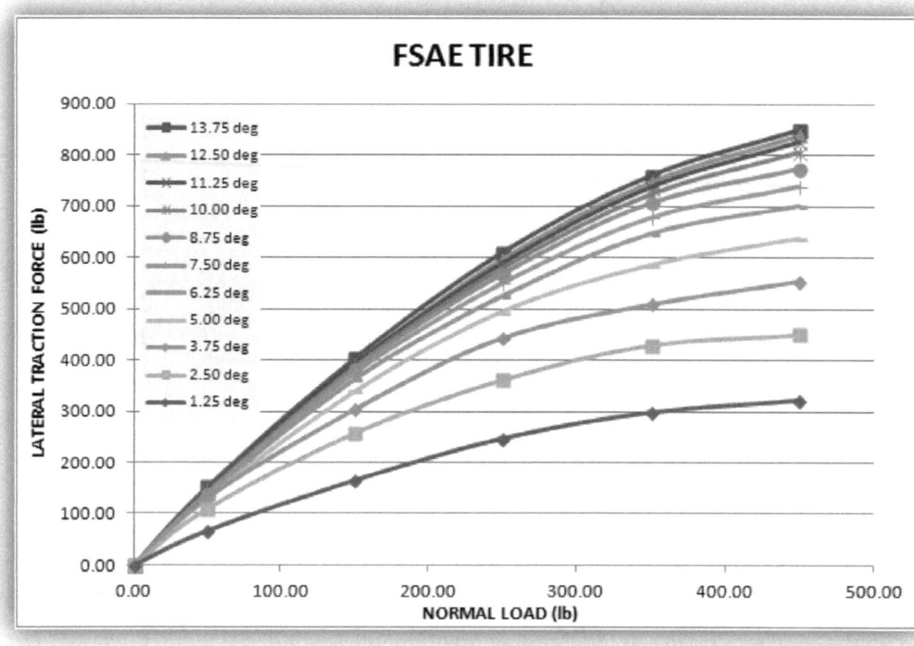

Figure 2.5 – TIRE NORMAL LOAD, LATERAL TRACTION, SLIP ANGLE RELATIONSHIP

A major problem with all simulations and studies based on linear models is that tire behavior beyond the proportional limit is not taken into account. Although road going passenger vehicles seldom venture into the nonlinear transition region between the proportional limit and the deformation limit, racing vehicles must always do so. A simulation/study of race car behavior would seem to require a non-linear curve fit to the entire "useful traction" (both "linear" and "transition" portions) region[41]. For passenger car tires the deformation limit tends to correspond to drift angles of around 12 degrees; race car tires may go higher. The great skill of racing drivers is the ability to "walk the tight rope" of the transition region without "falling off" (going into a skid).

To illustrate the lateral disturbance reaction possibilities, consider the case of three vehicles of differing longitudinal weight distribution, but with suspension and tire characteristics

[40] Reference [16], pg. 7. The exact tire used for the chart was not specified by the reference, but **FSAE** tires (special tires produced for use in the **SAE** Formula Student Design Competition) range in size from 18×6-10 to 20.5×7.5-13 and tend to run at inflation pressures of 12 to 16 psi on wheels of 7.0 in width.
[41] Reference [19], page 126.

all equal between vehicles, and equal fore and aft for each vehicle, encountering the same lateral disturbance force "F_d"[42]:

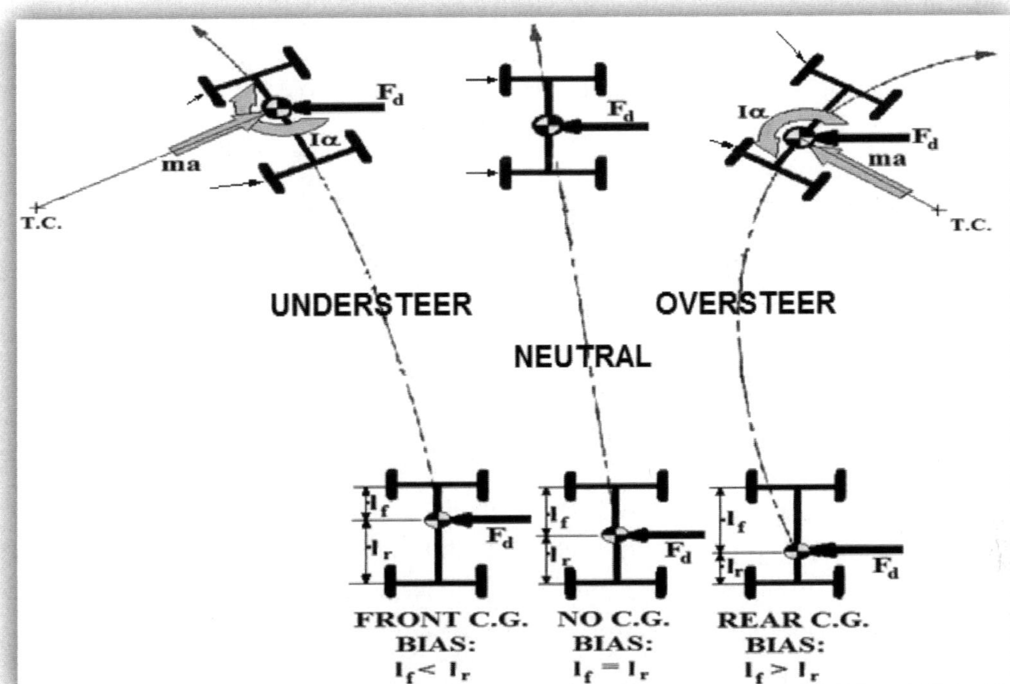

Figure 2.6 – THE THREE AUTOMOTIVE DIRCTIONAL STABILITY CASES

The forward **CG** bias case ("$l_f < l_r$") develops larger slip angles at the front tires then at the rear in response to the disturbance force "F_d"; this action results in the forward **CG** vehicle steering away from "F_d" generating an inertial reaction "$m\,a_y$" that tends to counteract the effect of the disturbance. Therefore this "**understeering**" action is considered to be "**stable**". The mid **CG** case ("$l_f = l_r$") develops equal slip angles all around in response to "F_d"; this results in no turning action, just a sideways shift, so no rotational inertial reaction is generated. The rear **CG** bias case ("$l_f > l_r$") develops larger slip angles at the rear tires than at the front in response to "F_d"; this results in a steering into the disturbance force "F_d" generating an inertial reaction "ma_y" that tends to augment the disturbance. This in turn causes an even sharper turning, generating an even larger inertial reaction. This sort of "**oversteering**" action is therefore considered to be "**unstable**" as it tends to "**feed upon itself**".

It was such an explanation that Olley developed, although with a more nuanced understanding, of over/under/neutral steering by 1934. The essence of automotive directional stability is inherent in this "explanation" as just given. However, this "explanation" falls far short of an authentic analysis in that it:

[42] Reference [28], pg. 23.Reference [3], pg. 10.

✓ **Does not show the variation in stability with speed.**
✓ **Does not include the effects of roll.**
✓ **Is only qualitative, not quantitative.**
✓ **Does not allow for variance front to rear in suspension and/or tires.**

An actual analysis that would include all of the above took a long time to develop as was related in Chapter 1, which was a very condensed history of automotive directional stability theory development. At the heart of the theory is that automotive directional stability ultimately depends on the relationship of the slip angles at the tires, front vs. rear. From that basic relationship the consequent steering geometry may be determined, along with a number of quantitative directional stability metrics.

One of the most general metrics of directional stability is a quantity called the "**Understeer Gradient**" (a.k.a. "**Understeer Coefficient**", although that is a misnomer), which is symbolized as "$\mathbf{K_{us}}$" and is expressed in radians or degrees per "**g**". "$\mathbf{K_{us}}$" presents a more general way of classifying a vehicle's degree of stability than comparing slip angle values "ψ_f" to "ψ_r" at a specific lateral acceleration value. If a vehicle is stable (**Understeering** or **US**) then "$\mathbf{K_{us} > 0}$", if neutral (**Neutral Steering** or **NS**) then "$\mathbf{K_{us} = 0}$", or if unstable (**Oversteering** or **OS**): "$\mathbf{K_{us} < 0}$". This metric can take various forms, of which the simplest is:

$$K_{us} = \left[\frac{W_f}{gC_{sf}} - \frac{W_r}{gC_{sr}}\right] \qquad \textbf{(EQ. 2.1)}$$

This equation comes about through that simple "comparison" of the front to rear slip angles at some lateral acceleration level as already discussed:

$$K_{us} \ @ \ a_y \approx \psi_f - \psi_r$$

This comparison becomes more general by substitution of the appropriate cornering stiffness expressions for the slip angles ($\psi = F_y/C_s$):

$$K_{us}@ \ a_y = \frac{F_{yf}}{C_{sf}} - \frac{F_{yr}}{C_{sr}}$$

The expressions for the front axle and rear axle lateral forces ("$\mathbf{W_f \, a_y/g}$" and "$\mathbf{W_r \, a_y/g}$") may be substituted in place of the symbols for those forces ("$\mathbf{F_{yf}}$" and "$\mathbf{F_{yr}}$"):

$$K_{us}@ \ a_y = \frac{W_f a_y}{gC_{sf}} - \frac{W_r a_y}{gC_{sr}}$$

Since the value of the lateral acceleration "$\mathbf{a_y}$" is the same on both sides of the equation it can be dropped, and the result is termed the Understeer Gradient ("$\mathbf{K_{us}}$") as shown before:

$$K_{us} = \left[\frac{W_f}{gC_{sf}} - \frac{W_r}{gC_{sr}}\right]$$

It is important to note that this metric applies most readily to a vehicle in a steady-state, not transient, condition. This metric covers what might be called "**static**" (quasi-static) yaw stability; it tells us nothing about "**dynamic**" (oscillatory) yaw stability, but usually modern vehicles have sufficient yaw damping to render such oscillations null (critical yaw damping) or a short diminishing series[43].

While the "**Understeer Gradient**" is the automotive stability (handling) metric (figure of merit) that will be central to the purpose of this paper, it is only one of about five classical stability metrics that were developed over the course of the theoretical development outlined in Chapter 1. Those five classic automotive directional stability metrics are[44]:

1. "**Understeer Gradient**" ($\mathbf{K_{us}}$) is a quantitative stability metric expressed in radians or degrees per "**g**".
2. "**Characteristic Speed**" ($\mathbf{V_{char}}$) is a metric, calculated at zero lateral acceleration, for comparing understeering designs with respect to the exact degree of understeer present.
3. "**Critical Speed**" ($\mathbf{V_{crit}}$) is a speed, calculated at zero lateral acceleration, where an oversteering vehicle can switch from initially controllable (slight oversteer) to very unstable (high oversteer) behavior.
4. "**Steady-State Steering Angle**" ($\boldsymbol{\delta_{ss}}$) steering angle required to make steady-state turn of radius "**R**" at "$\mathbf{a_y}$"; is an indicator of under/over steer and of steering responsiveness.
5. "**Static Margin**" (**SM**), is a quantitative stability metric like "$\mathbf{K_{us}}$", but is expressed as a dimensionless fraction. It constitutes a direct borrowing from aircraft stability theory, with traction forces substituted for aerodynamic forces.

---

[43] Reference [3], pp. 4 & 7. Rear suspension systems having adverse "**roll steer**" characteristics are stated as being known contributors to such oscillatory behavior; see Reference [12], pp. 220-221.

[44] Reference [19], pp. 161-164. Reference [12], pp. 203-208. There are a number of other significant stability metrics, plus variations in nomenclature/symbolism (such as **Under/Over Steer Gradient** "**U**" for **Understeer Gradient** "$\mathbf{K_{us}}$"). These variations on the metrics as presented herein result from analyses of varying sophistication and complexity. The result is a plethora of various "***stability metrics***", or "***figures of merit***", which are of differing usage and value. However, this paper will be limited in consideration to just the "**Understeer Gradient**" as defined herein.

3 – THE SKIDPAD SIMULATION

> **"Don't let anyone persuade you that the skid pad tests don't mean anything. They mean pretty much everything if we will just take the trouble to interpret them."**
>
> **Maurice Olley**[45]

A "**simulation**", as utilized in this paper, is a computer program that constitutes a transformation of a mathematical model(s) of some physical phenomena into a set of algorithms via utilization of some appropriate programming language. No model of physical reality is ever totally representative of reality, simplifications are inevitable due to ignorance, lack of time and/or resources, or inadvertence. All of the models ever made of physical reality are lacking in some way, and when significant deviations of theory from reality are noted then it is customary to revise the theoretical model, causing it to grow ever more complex and ever more in accord with reality. However, no matter how complex the model, it will still always be lacking in some way; the criterion oft used for deciding whether a model is sufficiently realistic, or requires further development, is whether the present state of the model reflects reality sufficiently to be useful[46].

The physical model reflected in the automotive skidpad simulation upon which this paper is based has numerous simplifications, some of which are known and will be identified in due course. There are also possible inadvertent and unknown simplifications due to the state of the underlying theory, limitations of the programming language and computer environment, and constraints on time and resources. However, the author of this paper believes that the skidpad simulation upon which this paper is based is not only sufficient for the paper's stated purposes, but with development may far exceed some of those great classical models which brought the theory of automotive directional stability to its present state. This is not due to any greater ability on the part of this author with regard to all those wondrous individuals who developed the theory, but due to the immense advantages this author has of having access to modern computer resources and to all that developed theory[47].

---

[45] Reference [20], pg 49. This quote was found in a letter from Olley to a colleague that was written sometime in the early 1930's.
[46] Reference [29], pg. 6.
[47] Sir Isaac Newton is credited with saying "***If I have seen further than others it is because I have stood on the shoulders of giants***." in 1675. However, French philosopher and scholar Bernard of Chartres is reputed to have said essentially the same thing sometime just prior 1129 AD...

It was during 1931-1932 that Maurice Olley and the Cadillac Research Group at the General Motors Proving Ground first conceived of, and then developed, the "**skidpad**" as a means for assessing automotive handling. However, most people think of a skidpad as merely a test course for determining a vehicle's maximum lateral acceleration capability. While a skidpad is commonly used for that purpose, it can be used for so much more.

When a vehicle is circling around a skidpad in a steady-state[48] condition, the steering angle held "constant", the vehicle velocity may be slowly increased (with steering increased equally slowly) and the vehicle response noted. If the vehicle attitude changes, taking a nose in/tail out position with respect to a tangent to the circle, and then changes course so as to spiral in toward the center of the circle ("**- - -**" path), then the vehicle is said to "**oversteer**". If, however, the vehicle adopts a nose out/tail in attitude with respect to a tangent to the circle, and changes course so as to spiral out away from the center of the circle ("**oooo**" path), then the vehicle is said to "**understeer**". If the vehicle tends to neither "**oversteer**" nor "**understeer**" but instead continues to circle the pad at a constant radius ("**—**" path), then the vehicle is said to be "**neutral**".

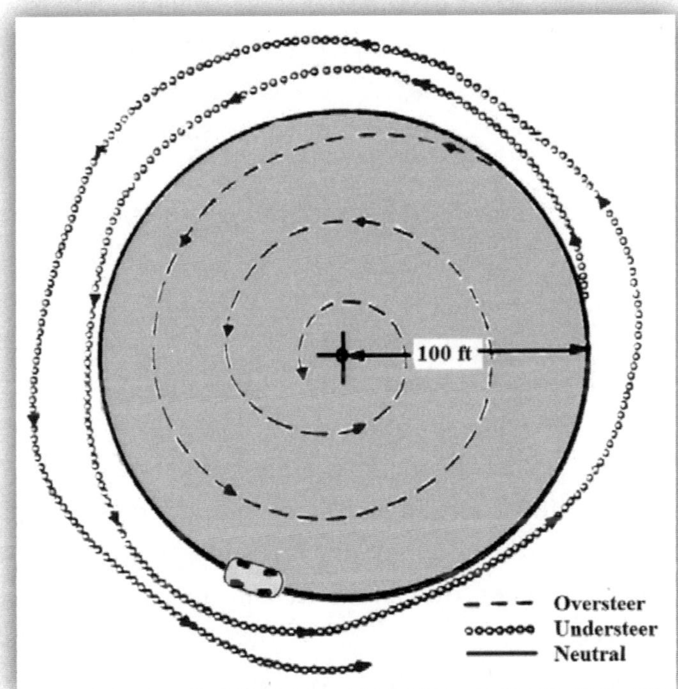

Figure 3.1 – SKIDPAD AND HANDLING CHARACTERISTICS

[48] The chief distinction between the steady-state and the transient condition is the absence of yaw acceleration ("**α**") in the steady-state. Along with "**α = 0**", in the steady-state there is no fluctuation in traction forces, no change in roll angle, no change in vehicle attitude, no change in vehicle yaw center (in the steady-state condition vehicles tend to yaw about the **CG**), etc.

The "**oversteer**"/"**understeer**"/"**neutral**" vehicle handling character is determined by a number of factors, but the most basic is the vehicle **longitudinal mass distribution**. The oversteering character is associated with an **aft weight bias**. The understeering character is associated with a **forward weight bias**. And, of course, the neutral handling character is associated with a vehicle **CG** near the midpoint of the wheelbase. However, these generalizations hold only for a conventional road vehicle with the same type/size tires all around, equal inflation pressures all round, and suspension characteristics reasonably the same at the front and rear.

The "trouble" with a neutral steering vehicle is that the neutral condition tends to not be constant; as a vehicle is operated fuel is consumed, loading can vary, and onboard objects can shift positions. All of this variation can result in a shifting **CG** location during operation, which in turn can cause the vehicle handling character to change, perhaps from neutral to understeer then back to neutral on the way to oversteer again. Such change in a vehicle's basic handling character can be disconcerting to the driver and therefore is inherently dangerous. For this reason it is generally best if a road vehicle maintains a degree of understeering character throughout its operational envelope. However, race car design may intentionally include an amount of directional instability for sharp low speed maneuvers, with a change to increased directional stability for sweeping high speed maneuvers; this has to do with the fastest way to get around a road race course with concession to the speed realm where stability is most critical.

For the skidpad there is no official standard construction, as may be seen from the following examples:

- General Motors historically used a 108 ft (32.9 m) radius skidpad.
- Road & Track magazine used a 100 ft (30.5 m) radius skidpad.

However, skidpad size is critical; the larger the radius the less angularity is involved in resolution of forces at the tires, and thus a higher maximum lateral acceleration may be obtained. Also, if the vehicle in question is designed to generate an aerodynamic "**down force**", then the higher velocities required on a larger radius pad to reach same lateral acceleration ("$\mathbf{a_y = V^2/R}$") results in greater "**down force**" and thus also to a higher maximum acceleration. The matter of "**down force**" is so significant that **Car and Driver** magazine recommended testing race cars such as the 1981 Lola T600, an **IMSA GTP** car, on a skidpad of 372 ft (113.4 m) radius, as opposed to a smaller 141 ft (43 m) radius pad. The Lola pulled 1.42 g's maximum lateral acceleration on the large pad, versus only 1.23 g's on the small one. How well the Lola would have done on the large pad without "**down force**" is open to question, but it would still have bested its small pad figure[49].

[49] Reference [32], pg. 30.

The point is that empirically obtained lateral acceleration values depend on exactly how they were obtained, and the size of the skidpad is the most basic consideration. Full information regarding skidpad test conditions is seldom given. For instance, what was the material (asphalt, concrete) of the skidpad, was the surface clean or dusty, damp or dry. Also, how was the test conducted; are the results an average of runs going in both directions[50]? And, what was the condition of the vehicle at the time of test; was the vehicle in some standard and reproducible configuration[51]?

As a prerequisite for a skidpad simulation a reliable and accurate means of tire data handling must be established. Instead of using purchased tire data (which isn't available in this case anyway) formatted for use in a commercial tire modeling program (probably "**Magic Formula**" based), a measured data "**pick-off**" from each of the slip angle family of curves was made for each constant normal load in the following tire chart, which is a chart for the 6.00×16 bias ply as utilized by the 1958 Jaguar XK150S[52]. For each normal load a "**ψ = f(F_y)**" relationship was determined by a subsequent "**curve fitting**" (via "**regression analysis**"[53]) to that "**pick-off**" data.

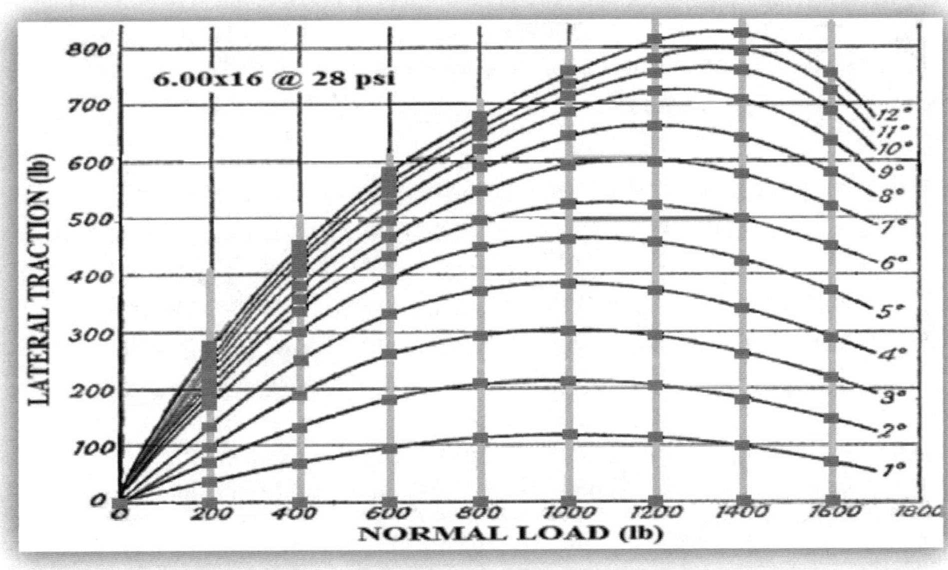

Figure 3.2 – TIRE DATA "PICK-OFF"

[50] Most road vehicles are initially balanced near centerline, which means that in test configuration with a driver aboard the **CG** is slightly biased laterally toward the driver. This means a left side steering car will do best turning left, and a right hand steering car will do best turning right. The 2010 Dodge Viper had an engine offset of 15 mm toward the passenger side so as to offer maximum uniformity in behavior whether turning left or right when running in the most sporting mode of "**driver only**".

[51] On at least one occasion after skidpad testing a test vehicle was found to hold many pounds of extraneous equipment in the trunk, residue from previous testing activities.

[52] Reference [8], pg. 165.

[53] Regression analysis has long been a "weapon" in the mass property engineer's "arsenal", although it is most commonly used for development of **WER's** (**Weight Estimation Routines**). For more information on regression analysis see Reference [31], pp. 3-7, Reference [30], pp. 19-39, and Reference [36], pg. 16.7.

The following is an example of the process: the "**pick-off**" data shown below ("**FOR N = 400**" lb) is subjected to a regression analysis such that an equation "$\psi = f(F_y)$ @ **N**" is obtained. That equation is then be subjected to an inverse first order differentiation "$\partial\psi/\partial F_y$" such that a cornering stiffness equation "$C_s = (\partial\psi/\partial F_y)^{-1}$ @ **N**" is obtained. The cornering stiffness is a partial differential because the slip angle doesn't vary just as a function of lateral load, but also as a function of normal load, inflation pressure, temperature, and velocity[54]. We are holding all these other variables constant, which is the essence of taking a partial differential.

FOR N = 400:

F_y	ψ
0	0
69.97587	1
130.9204	2
190.6239	3
250.8101	4
301.758	5
337.1251	6
366.7394	7
379.1796	8
399.8621	9
419.2347	10
432.6784	11
446.191	12

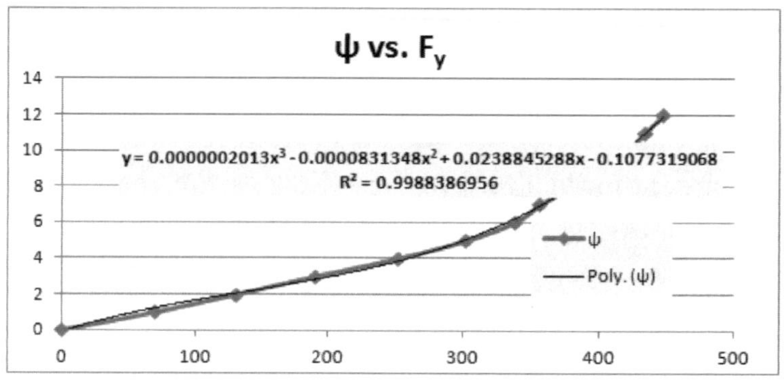

$$\psi = 0.0000002013\,F_y^{\,3} - 0.0000831348\,F_y^{\,2} + 0.0238845288\,F_y - 0.1077319068$$

$$C_s = (0.0000006039\,F_y^{\,2} - 0.0001662696\,F_y + 0.0238845288)^{-1}$$

Figure 3.3 – TIRE DATA REGRESSION ANALYSIS EXAMPLE

Now what is needed is an interpolation routine such that for any lateral force value not falling directly on a constant normal load line (which would likely be all the lateral force values in the simulation) the appropriate slip angle and cornering stiffness values can still be obtained[55].

Consider the interpolation example: to find the slip angle at outer front tire when "$\mathbf{a_y} = \mathbf{0.06}$ **g's**, $\mathbf{F_{yaof} = 56.4}$ **lb**, and $\mathbf{N_{of} = 971.6}$ **lb**"; first we "bracket" the actual normal load by evaluating the slip angle @ "$\mathbf{F_{yaof} = 56.4}$ **lb**" using slip angle functions for normal loads just lesser and greater than the "$\mathbf{N_{of} = 971.6}$ **lb**":

$$\mathbf{N = 800\ lb,\ \psi_{800} = 0.63\ deg}$$

$$\mathbf{N = 1000\ lb,\ \psi_{1000} = 0.57\ deg}$$

Assuming a short interval (200 lb) linear variation of slip angle with normal load[56] the outer front tire slip angle "ψ_{of}" at a normal load of "$\mathbf{N = 971.6}$ **lb**" may now be approximated:

---

[54] For this paper's simulation it is assumed that a condition of thermal equilibrium has been attained such that temperature and pressure are constant; the variation of traction coefficient with velocity is taken as negligible.
[55] Reference [33], pp. 99-100.

$$\psi_{of} = \left(\frac{\psi_{1000} - \psi_{800}}{1000 - 800}\right) \times (971.6 - 800) + \psi_{800} = 0.58\ degrees$$

The preceding slip angle approximation process may be carried out for each tire on the vehicle model. The cornering stiffness values at each tire may be found by a similar interpolation of the cornering stiffness equations at constant normal load which "**bracket**" the actual normal load values.

Also needed is a data adjustment routine such that the slip angle "ψ" and cornering stiffness "C_s" results will be valid for inflation pressures other than that of the original tire data. Presented here is an example of the data adjustment routine such that the slip angle "ψ" and cornering stiffness "C_s" function outputs will be valid for inflation pressures (such as 40 psi) other than that of the original data (which in this case was 28 psi); note that the revised cornering stiffness must be determined first[57]:

$$Cs_{40\,psi} = Cs_{28\,psi}\left(\frac{Kz_{40\,psi}}{Kz_{28\,psi}}\right)$$

$$\psi_{40\,psi} = \psi_{28\,psi}\left(\frac{Cs_{28\,psi}}{Cs_{40\,psi}}\right)$$

The tire vertical stiffness values "K_z" needed to commence the data adjustment are calculated per the Rhyne Equation[58]:

$$K_Z = 0.00028\, P_i \sqrt{(-0.004\, AR + 1.03) S_N \times \left[\frac{S_N \times AR}{50} + D_R\right]} + 3.45 \quad \textbf{(EQ. 3.01)}$$

Where:

P_i = Tire inflation pressure (kPa).

K_Z = Tire vertical stiffness (kg/mm).

AR = Tire section aspect ratio (dimensionless).

S_N = Tire nominal section width (mm).

D_R = Wheel rim nominal diameter (mm).

---

[56] This interpolation process can be considerably improved by assuming a nonlinear variation, which is certainly the physical case, and/or by developing more closely spaced (say, at every 100 lb increment in normal load) slip angle at constant normal load equations.

[57] Reference [33], pp. 101-102.

[58] Reference [11], pg. 192. The assumption here is that the tire lateral stiffness will vary in direct proportion to the tire vertical stiffness.

With the tire data handling routines now firmly established, the actual flow of the skidpad simulation, driven by incrementation of the lateral acceleration in g's (**$a_y = V^2/gR$**), may be examined step by step. The overall relation of these steps to each other is illustrated by the following flow chart:

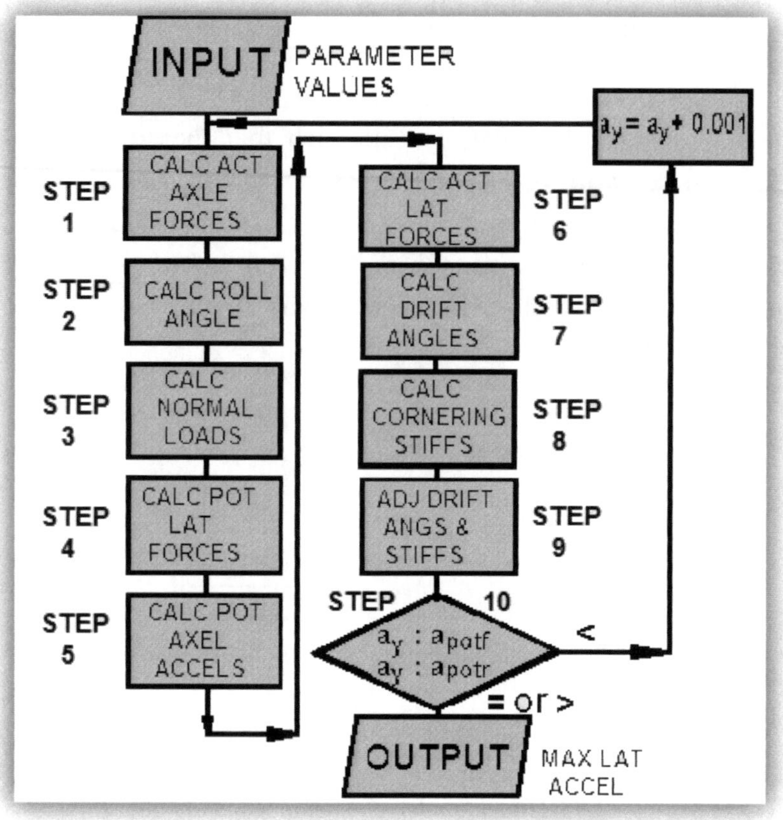

Figure 3.4 – SKIDPAD SIMULATION

This iterative method can produce a reasonably good estimate of a vehicle's maximum lateral acceleration, but also can reveal a great deal of other valuable info, such as the slip angles at each tire (**ψ_{fi}, ψ_{fo}, ψ_{ri}, ψ_{ro}**), the inner and outer steering angles (**δ_i, δ_o**), the "**vehicle sideslip angle**" (**β**)[59], and directional stability metrics such as "**K_{us}**", at each lateral acceleration value. The simulation can exceed the ability of a physical skidpad in one respect; if the normal termination of the program at maximum lateral acceleration is overridden, then a value for the vehicle rollover lateral acceleration can also be determined.

To implement this method of skidpad simulation a spreadsheet (**MS-Excel**) was set-up as it was quicker to do so than to write a program in either **GWBASIC** or **FORTRAN**. For development and "**validation**" runs on this spreadsheet a 1958 Jaguar XK150S was used as the test subject. This vehicle was chosen as the simulation subject because it was the only vehicle for which a full set of parametric data was readily available.

---

[59] This is the preferred **SAE** terminology for "**β**", along with "**attitude angle**" as a secondary possibility (see Reference [35], pg. 17.

The iterative method as used for the skidpad simulation in this paper is basically the same as that presented by Clemson University professors Karogal, Law, and Ayalew in their paper "**An Iterative Approach for Steady State Handling Analysis of Vehicles**". The major difference between the method proposed by the Clemson professors and method used in this paper is that the Clemson method utilized commercial **Pacejka** ("**Magic Formula**") software to produce tire data, instead of data pick-off from a tire chart, to produce a tire model via regression analysis[60]. The simulation used by this paper may now be examined step by step…

Step 1, the required (actual) lateral forces at the front and rear axles ($\mathbf{F_f}$,$\mathbf{F_r}$) are determined by multiplying the front and rear axle static **weight** loads ($\mathbf{W_f}$,$\mathbf{W_r}$) by the lateral acceleration "$\mathbf{a_y}$"[61]:

$$F_f = W_f a_y \qquad \textbf{(EQ. 3.02)}$$

$$F_r = W_r a_y \qquad \textbf{(EQ. 3.03)}$$

Step 2, the sprung mass roll angle "$\boldsymbol{\theta_s}$" due to the lateral acceleration "$\mathbf{a_y}$" may be determined by use of the vehicle's "**roll gain**" ("**RG**") value[62]:

$$\theta_s = RG \times a_y \qquad \textbf{(EQ. 3.04)}$$

Step 3, the normal loads on each tire ($\mathbf{N_{of}}$, $\mathbf{N_{if}}$, $\mathbf{N_{or}}$, $\mathbf{N_{ir}}$) in response to the lateral acceleration ("$\mathbf{a_y}$") induced load transfer are now determined by the equations[63]:

$$N_{of} = \frac{W_f}{2} + \frac{12\theta_s K_{froll}}{t_f} + W_s a_y \frac{l_{rs}}{l_{wb}} \frac{h_{rcf}}{t_f} \qquad \textbf{(EQ. 3.05)}$$

$$N_{if} = \frac{W_f}{2} - \frac{12\theta_s K_{froll}}{t_f} - W_s a_y \frac{l_{rs}}{l_{wb}} \frac{h_{rcf}}{t_f} \qquad \textbf{(EQ. 3.06)}$$

$$N_{or} = \frac{W_r}{2} + \frac{12\theta_s K_{rroll}}{t_r} + W_s a_y \frac{l_{fs}}{l_{wb}} \frac{h_{rcr}}{t_r} \qquad \textbf{(EQ. 3.07)}$$

$$N_{ir} = \frac{W_r}{2} - \frac{12\theta_s K_{rroll}}{t_r} - W_s a_y \frac{l_{fs}}{l_{wb}} \frac{h_{rcr}}{t_r} \qquad \textbf{(EQ. 3.08)}$$

---

[60] Reference [15]. Another big difference is that the Clemson professors had student assistants to do all the "**grunt work**" to produce their paper. More important from a technical standpoint, they were also able to generate tire camber thrust and aligning torque equations this way as well.

[61] Reference [33], pg.104.

[62] Reference [32], pg. 21. A "typical" roll gain value might be something like "**8.6 deg/g**". Roll gains are linear approximations of steady-state equilibrium roll functions, typically of a form such as "$\boldsymbol{\theta_s = A\, a_y^2 + B\, a_y + C}$", which may be used in lieu of "**RG**".

[63] Reference [22], pp. 93-94. Reference [19], pg. 683.

Step 4, given the normal loads at each tire, the lateral tire traction force potentials (**Fy_{pof}, Fy_{pif}, Fy_{por}, Fy_{pir}**) at the lateral acceleration "**a_y**" can now be found[64]:

$$F_{ypof} = (b_{of} - m_{of}N_{of})N_{of} \quad \textbf{(EQ. 3.09)}$$

$$F_{ypif} = (b_{if} - m_{if}N_{if})N_{if} \quad \textbf{(EQ. 3.10)}$$

$$F_{ypor} = (b_{or} - m_{or}N_{or})N_{or} \quad \textbf{(EQ. 3.11)}$$

$$F_{ypir} = (b_{ir} - m_{ir}N_{ir})N_{ir} \quad \textbf{(EQ. 3.12)}$$

Step 5, given these lateral traction force potentials, the potential lateral acceleration at each axle (**a_{yfpot}, a_{yrpot}**) can now also be found per[65]:

$$a_{yfpot} = (F_{ypof} + F_{ypif})/W_f \quad \textbf{(EQ. 3.13)}$$

$$a_{yrpot} = (F_{ypor} + F_{ypir})/W_r \quad \textbf{(EQ. 3.14)}$$

Step 6, is based on the assumption that actual lateral traction forces at each tire (**F_{yaof}, F_{yaif}, F_{yaor}, F_{yair}**) will be in proportion to the potential lateral traction force at that tire (**F_{ypof}, F_{ypif}, F_{ypor}, F_{ypir}**) times the acceleration ratio "**a_y/a_{yfpot}**" or "**a_y/a_{yrpot}**" as appropriate. This assumption may not hold very well at lateral acceleration levels far short of the vehicle's maximum lateral acceleration capability, but at lateral acceleration levels nearing tire traction force saturation it should be just fine. Anyway, the assumption means the actual lateral force at each tire is to be determined by[66]:

$$F_{yaof} = F_{ypof}\left(\frac{a_y}{a_{yfpot}}\right) \quad \textbf{(EQ. 3.15)}$$

$$F_{yaif} = F_{ypif}\left(\frac{a_y}{a_{yfpot}}\right) \quad \textbf{(EQ. 3.16)}$$

$$F_{yaor} = F_{ypor}\left(\frac{a_y}{a_{yrpot}}\right) \quad \textbf{(EQ. 3.17)}$$

$$F_{yair} = F_{ypir}\left(\frac{a_y}{a_{yrpot}}\right) \quad \textbf{(EQ. 3.18)}$$

Step 7, now the actual lateral force at each tire is known, the slip angle at each tire may be found by using the regression analysis slip angle equations that "**bracket**" the actual normal

---

[64] Reference [17], pg. 27. Reference [18], pg. 127. The coefficients "**b**" and "**m**" are tire characteristics.
[65] Reference [33], pg. 107.
[66] Ibid, pg. 108.

load so that the slip angle may be determined by a linear interpolation process as per the following example.

For this example "**a_y = 0.06 g's, N_{of} = 971.6 lb** and **F_{yaof} = 56.4 lb**"; the slip angle equations which "**bracket**" the "**N_{of} = 971.6 lb**" normal load are the equations for constant normal load values "**N=800**" and "**N=1000 lb**":

$$\psi_{800} = 0.0000000520\ F_y^3 - 0.0000321034\ F_y^2 + 0.0152816867\ F_y - 0.1398755826$$

$$\psi_{1000} = 0.0000000314\ F_y^3 - 0.0000214695\ F_y^2 + 0.0139331283\ F_y - 0.1532355494$$

These equations are evaluated at "**F_y = 56.4 lb**", resulting in "**ψ_{800} = 0.63 deg**" and "**ψ_{1000} = 0.57 deg**". Now the actual slip angle "**ψ_{of} for N_{of} = 971.6 lb**" may be approximated by the linear interpolation:

$$\psi_{of} = \left(\frac{0.57 - 0.63}{1000 - 800}\right) \times (971.6 - 800) + 0.63 = 0.58\ deg$$

<u>Step 8</u>, the cornering stiffness at each tire may also be determined by the same sort of linear interpolation process using the regression analysis cornering stiffness equations at constant normal load that "**bracket**" the actual normal load, which for the example "**a_y = 0.06 g's, N_{of} = 971.6 lb**, and **F_{yaof} = 56.4 lb**" are the cornering stiffness equations for "**N=800**" and "**N=1000 lb**":

$$C_s = (0.000000156\ F_y^2 - 0.0000642068\ F_y + 0.0152816867)^{-1} \text{ for } N = 800 \text{ lb}$$

$$C_s = (0.0000000942\ F_y^2 - 0.000042939\ F_y + 0.0139331283)^{-1} \text{ for } N = 1000 \text{ lb}$$

These equations are evaluated at "**F_y = 56.4 lb**", resulting in "**C_{S800} = 82.26 lb/deg**" and "**C_{S1000} = 84.67 lb/deg**". Now the cornering stiffness "**Cs_{of} for N_{of} = 971.6 lb**" may be approximated by the linear interpolation:

$$C_{Sof} = \left(\frac{84.67 - 82.26}{1000 - 800}\right) \times (971.6 - 800) + 82.26 = 84.33\ \text{lb/deg}$$

<u>Step 9</u>, at this point the slip angle and cornering stiffness at each tire must be adjusted if the inflation pressure of the tire data used for regression analysis is different than the inflation pressure of the lateral acceleration simulation.

For example: the regression analysis formulae were based on tire data measured at 28 psi inflation pressure, but the vehicle simulation is at 40 psi front inflation pressure (rear inflation pressure is 45 psi[67]), so adjustment of the cornering stiffness output is required. For this example "**a_y = 0.06 g's, N_{of} = 971.6 lb** and **F_{yof} = 56.4 lb**", so the adjustment formulae are:

$$C_{s\,40psi} = C_{s\,28psi}(K_{z\,40\,psi}/K_{z\,28psi})$$

$$\psi_{40\,psi} = \psi_{28\,psi}(C_{s\,28psi}/C_{s\,40psi})$$

When evaluated for this example case these formulae, and the resultant values, become:

$$\mathbf{C_{sof40} = 84.34\ (1331.0\ /\ 988.0) = \underline{113.62}\ \underline{lb/deg}}$$

$$\mathbf{\Psi_{of40} = 0.58\ (84.34/113.62) = \underline{0.43}\ \underline{deg}}$$

Step10, this is the point where the skidpad simulation will either terminate (because the max lateral acceleration has been reached: the actual lateral acceleration ("$\mathbf{a_y}$") equals the lateral acceleration potential ("$\mathbf{ay_{fpot}}$" or "$\mathbf{ay_{rpot}}$") at one of the axles) or the actual lateral acceleration will be incremented by a discrete amount (say, "**$a_y = a_y$ + 0.001 g's**") and the program will loop around back to Step 1 for a repeat of the whole process.

A "**validation**"[68] run of the skidpad spreadsheet utilized 1958 Jaguar XK150S parameter values. A maximum lateral acceleration ("$\mathbf{ay_{max}}$") of **0.67 g's** resulted; run termination was by terminal oversteer (Jaguar test **weight distribution** was 48%/52%); with manual over-ride of the maximum lateral acceleration run termination a rollover lateral acceleration ("$\mathbf{ay_{rollover}}$") of **0.89 g's** was attained.

At a selected value of 0.384 g's the 100 foot radius skidpad simulation slip angle output (ψ_{if}, ψ_{of}, ψ_{ir}, ψ_{or}) resulted in the Jaguar steering angles (δ_i, δ_o) and vehicle "**attitude angle**" (**β**)[69] as depicted in the following figure:

[67] Jaguar designers were apparently afraid that the XK150S would prove to be insufficiently understeering (stable); the specification of such a front-to-rear inflation pressure differential is a bit of a panic move, but not unknown.

[68] This does not constitute a true validation of the spreadsheet, but a properly rigorous validation is beyond the resources of the author. That the spreadsheet results so far seem reasonable gives a certain confidence.

[69] Reference [12], pp. 206-208. Reference [36], pg. 4.2.2.1.8. This author has been known to call "**β**" the vehicle "**Yaw Angle**", but official **SAE J670_2008** terminology is vehicle "**Side Slip Angle**" (which would seem to be, once again, a misnomer), or vehicle "**Attitude Angle**" (which seems preferable to vehicle "**Side Slip Angle**").

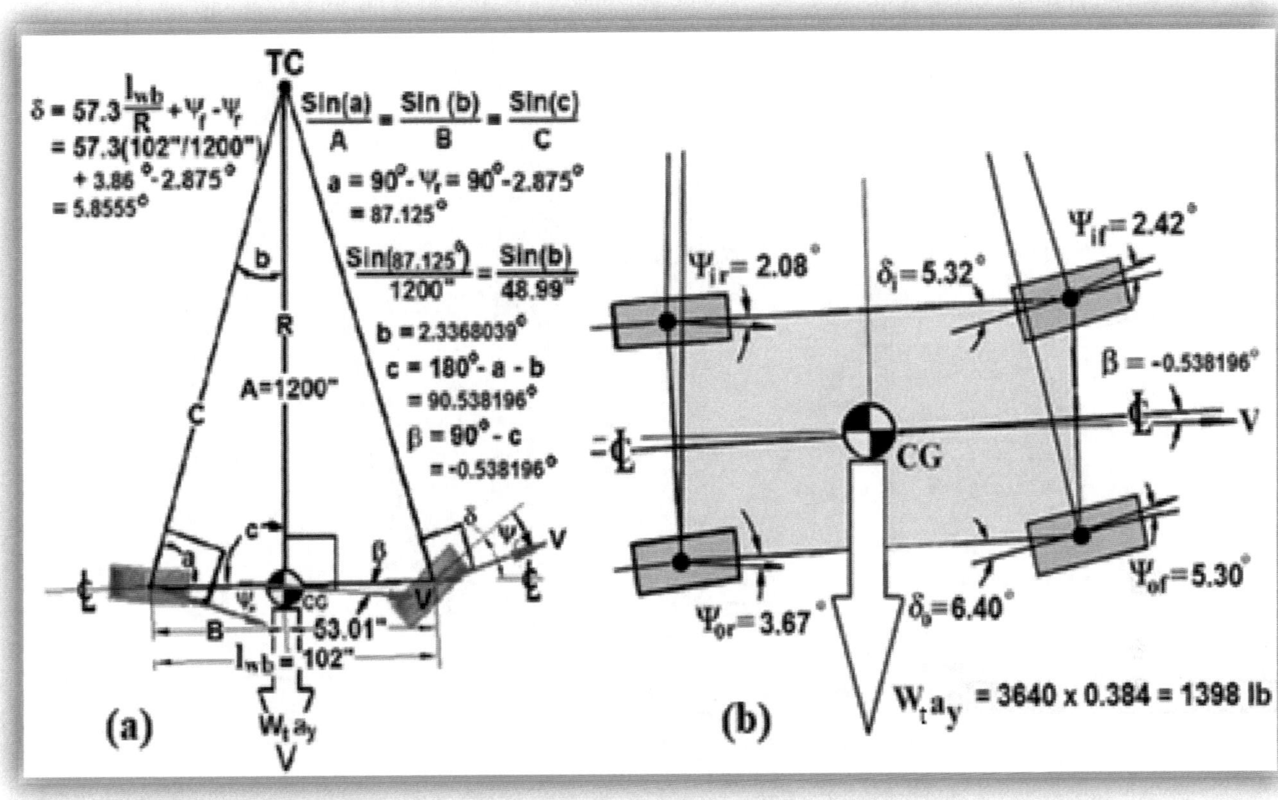

Figure 3.5 – JAGUAR XK150S CORNERING ATTITUDE @ 0.384 G's

The **SAE** sign convention states that the "**β**" angle is considered negative when the vehicle velocity vector "**V**" (through the **CG**) lies to the "outside" (away from the turn center, in the centrifugal force direction) of the vehicle centerline[70]. Therefore, the Jaguar in Figure 3.5 should have a negative attitude angle "**β**", which is in accord with the calculated value. All of the values shown in the Jaguar cornering attitude depiction of Figure 3.5 resulted from input of the vehicle (**l_{wb}, l_r**) and skidpad (**R**) parameters, plus the spreadsheet slip angle output (**ψ_{if}, ψ_{of}, ψ_{ir}, ψ_{or}**), into the established bicycle model relationships for "high speed" Ackermann steering geometry as per the following figure[71]:

[70] Reference [35], pg. 17 (Figure 3).

[71] Reference [19], pp. 126-128. Reference [20], pp.74-76. Note that the trigonometric relationship known as the "**Law of Sines**" can also play a part in defining the turning geometry (see Figure 3.5a). A four-wheeled illustration of the turning situation can be determined by treating the matter as two bicycle models side-by-side; note that the steering and front-rear slip angles of Figure 3.5a are merely the average of the corresponding angles of Figure 3.5b.

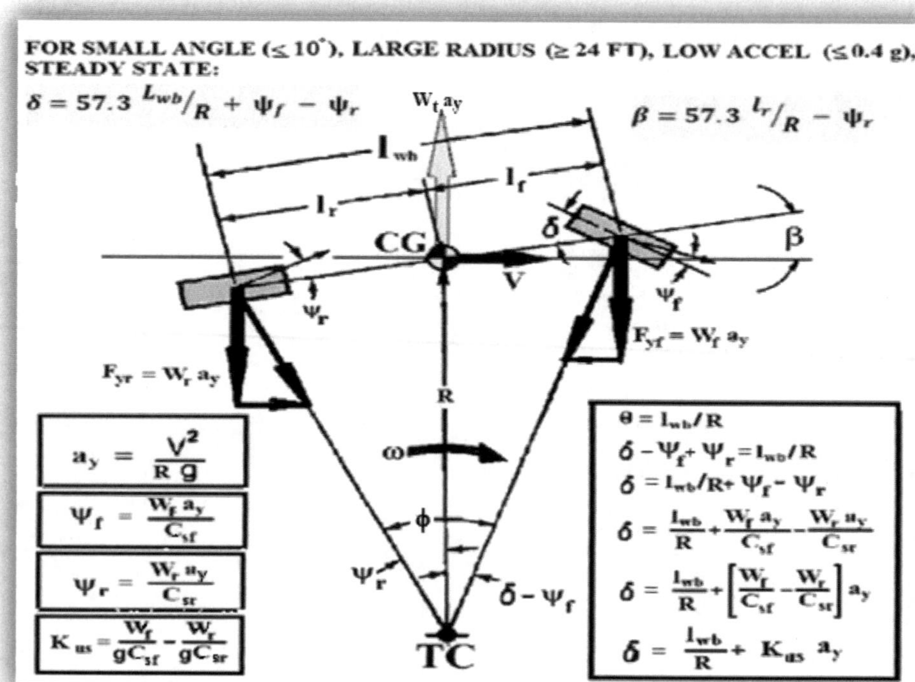

Figure 3.6 – BICYCLE MODEL RELATIONSHIPS

The "**validation**" run output of the simulation spreadsheet also allows for an Understeer Gradient (**K_{us}**), as well as other metrics, determination at selected lateral acceleration levels. When plotted as "**K_{us} vs. a_y**" the initial result for the XK150S was as depicted in Figure 3.7[72]:

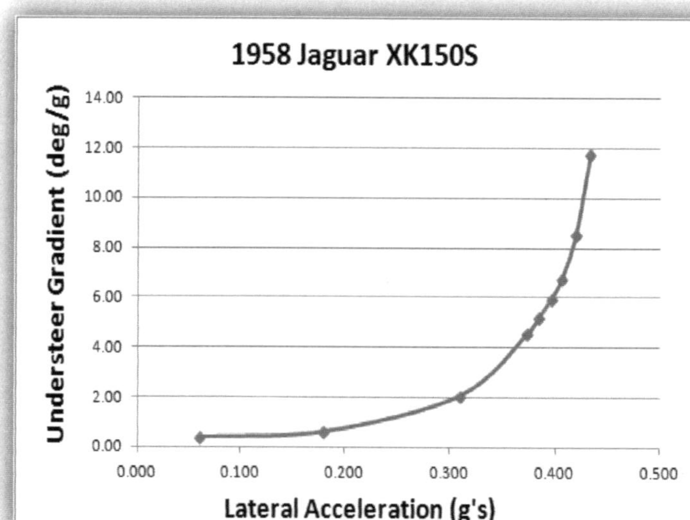

Figure 3.7 – JAGUAR XK150S UNDERSTEER GRADIENT PLOT

72 This was the result with the initial front suspension model; subsequent attempts to refine the front suspension model have yielded different results. What is needed is a validated model obtained from engineering drawings or measurements taken from an actual physical example.

Note that the form and magnitude of the Jaguar Understeer Gradient plot is not too different from what one would expect given the corresponding "typical" Understeer Gradient plots of Figure 3.8[73] (which is yet another small reason for some confidence in the simulation):

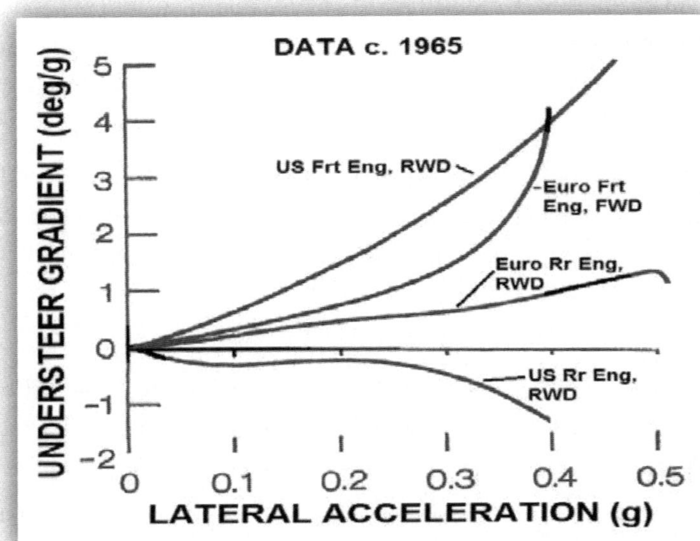

Figure 3.8 – TYPICAL UNDERSTEER GRADIENT PLOTS

[73] Reference [34], pg. 375.

4 – PARAMETER COUPLING

The use of a computer simulation instead of a physical model to explore some natural behavior has numerous advantages. That the use of a computer model may be much faster and economical than the use of a physical model is obvious and well understood, but less obvious is the fact that a computer simulation allows for an "**uncoupling**" of the relevant parameters affecting the behavior being explored. In the physical world to vary the **weight** of an automobile to study the effect upon performance must also mean some possibly unintended but accompanying changes in the **CG**, **roll/pitch/yaw inertias**, **sprung-to-unsprung weight ratio**, etc. In a computer simulation the vehicle **weight** can be changed and the effect noted without change to any other parameters, including the **roll/pitch/yaw inertias** (!!!).

This computer simulation ability to "**uncouple**" parameters can therefore lead to some very unrealistic conclusions if one is not careful. For instance, a change in vehicle **weight** without a corresponding change in **rotational inertias** might lead to totally erroneous conclusions given the <u>fundamental</u> nature of that "**connection**". Fortunately, for a steady-state skidpad simulation there are no angular accelerations, as would be in a transient simulation; the **roll/pitch/yaw inertias** play no role. However, the **weight** and other parameters may still play roles that affect the performance aspect under study in various, and sometimes hidden, ways. Therefore, it's best to make all the possible ways a parameter might affect the simulation explicit so as to avoid error[74]. This explicit understanding of the roles the various parameters play in this paper's steady-state skidpad simulation can only come about by a consideration of all the fundamental underlying relationships leading up to the simulated results. This would involve the roll angle equation (Eq. 4.1), the vehicle **CG** and roll moment arm geometry (Fig. 4.1, "Side"), roll stiffness (Eq. 4.2-4.3), and the basic **mass property** relationships (Eqs. 4.4-4.7). The roll angle equation is as follows[75]:

$$\theta_s K_{roll} = W_s a_y h_r \cos\theta_s + W_s h_r \sin\theta_s + W_{usf} a_y (r_{rf} - h_{rcf}) + W_{usr} a_y (r_{rr} - h_{rcr}) \quad \text{(EQ. 4.1)}$$

Where the symbolism is:

θ_s = The **sprung mass** roll angle (degrees).

K_{roll} = The total vehicle roll stiffness (lb-ft/deg).

W_s = The **weight** of the **sprung mass** (lb).

a_y = The lateral acceleration (g's).

h_r = The **sprung mass** roll moment arm (ft).

---

[74] Had the skidpad simulation been written differently, such that all "**primary**" parametric values could be input with automatic update of "**secondary**" or dependent parametric values (such as the roll gain "**RG**" or the front/rear **static weight** axle loads "**Wf, Wr**"), then such caution would not be so necessary. However, at the time the skidpad simulation was written massive parametric variation generating reams of data was not envisioned.

[75] Reference [10], pp. 133-136.

W_{usf} = The front axle **unsprung mass weight** (lb).

r_{rf} = The front axle **unsprung mass VCG** (approx. the rolling radius) (ft).

W_{usr} = The rear axle **unsprung mass weight** (lb).

r_{rr} = The rear axle **unsprung mass VCG** (approx. the rolling radius) (ft).

h_{rcf} = The distance of front roll center above (+) or below ground (-) (ft).

h_{rcr} = The distance of rear roll center above (+) or below ground (-) (ft).

The vehicle **CG** and roll moment arm geometry (from which "**h_r**" is determined for the above equation) is depicted[76]:

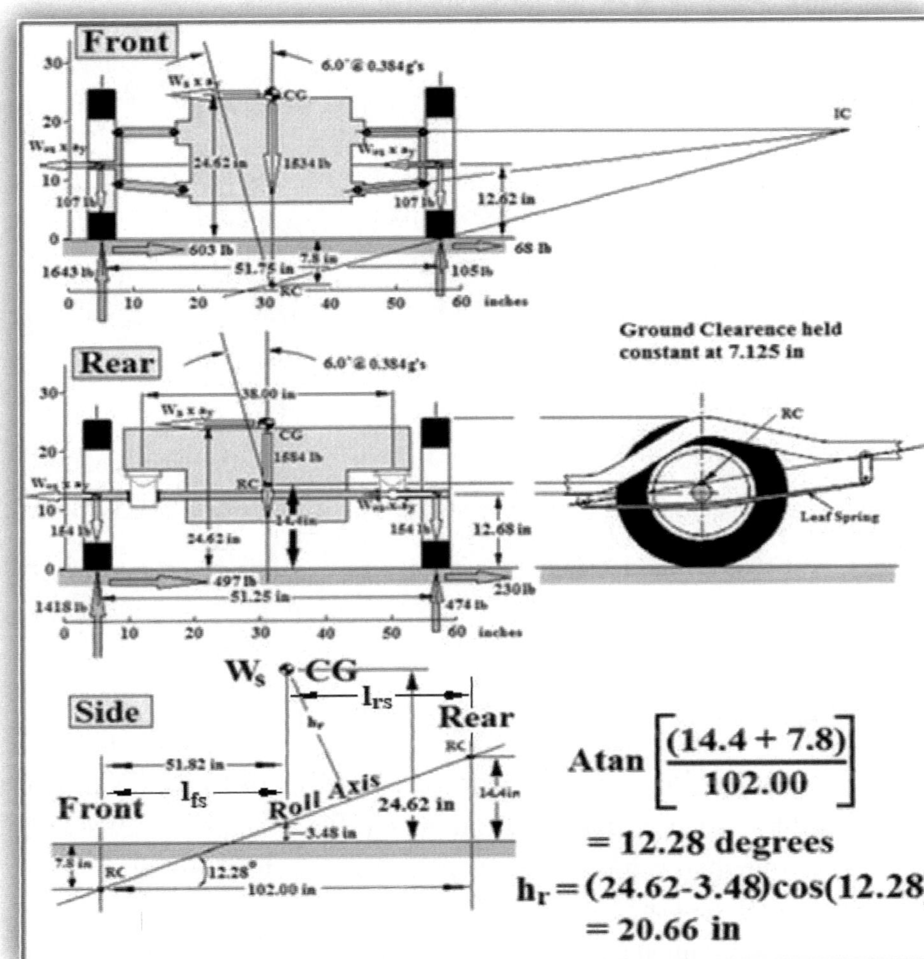

Figure 4.1 – ROLL AXIS GEOMETRY

While "**h_{rcf}**", "**h_{rcr}**", and "**h_r**" values used in Eq. 4.1 are determined from the geometry, the value of "**K_{roll}**" used in Eq. 4.1 is determined from the roll stiffness equations. For an

[76] Reference [33], pg. 61.

independent suspension (such as the front suspensions of most road vehicles) the roll stiffness equation is:

$$k_{roll} = \frac{k_{cs}\, t^2}{1375} \qquad \text{(EQ. 4.2)}$$

Where the symbolism is:

$\mathbf{k_{roll}}$ = Independent suspension roll stiffness (lb-ft/degree).

$\mathbf{k_{cs}}$ = Combined stiffness of suspension spring at wheel + tire (lb/in).

t = The track width at the axle line (in).

For a non-independent suspension (such as the rear suspension of many road vehicles) the roll stiffness equation is:

$$k_{roll} = \frac{(k_s t_{sb}{}^2)(k_t t^2)}{1375\,(k_s t_{sb}{}^2 + k_t t^2)} \qquad \text{(EQ. 4.3)}$$

Where the symbolism is:

$\mathbf{k_{roll}}$ = Non-independent suspension roll stiffness (lb-ft/deg).

$\mathbf{k_s}$ = Effective suspension spring stiffness at the mount to axle (lb/in).

$\mathbf{t_{sb}}$ = "Spring base" distance between spring mounts along an axle (in).

$\mathbf{k_t}$ = The spring stiffness of a tire on the axle (lb/in).

t = The track width at the axle (in).

To the roll stiffness as calculated at any one axle line the roll stiffness of any anti-roll bar (**ARB**) at that axle must be added in. Then the complete roll stiffnesses at the axle lines must be added together to obtain the total vehicle roll stiffness "$\mathbf{K_{roll}}$" as used in Eq. 4.1

A final word about parameter coupling involves some of the most significant parameters of all, and especially noteworthy ones given this paper's emphasis on **mass properties**. The **mass properties** of the subject vehicle used for this study, the 1958 Jaguar XK150S, were varied in accord with the usual "**Weight Accounting**" equations[77]:

[77] Reference [30], pg. 3.Reference [36], pg. 2.19. Only the first four of the ten "**Weight Accounting**" equations are relevant to this paper.

$$W_t = \sum W_i \qquad \text{(EQ. 4.4)}$$

$$\bar{X} = \sum W_i X_i / W_t \qquad \text{(EQ. 4.5)}$$

$$\bar{Y} = \sum W_i Y_i / W_t \qquad \text{(EQ. 4.6)}$$

$$\bar{Z} = \sum W_i Z_i / W_t \qquad \text{(EQ. 4.7)}$$

The **mass properties** for the subject vehicle in the nominal "**test configuration**"[78] were determined to be as per the following table (note that the **mass moments of inertia** are not included in the accounting as the "**steady-state**" condition of the simulation renders them superfluous):

Component	Symbol	Wt (lb)	LCG (in)	TCG (in)	VCG (in)
Sprung Weight	W_s	3117.93	51.82	0.00	24.62
Unsprung Weight Fwd	W_{fus}	214.03	0.00	0.00	12.62
Unsprung Weight Rr	W_{rus}	307.85	102.00	0.00	12.68
Total Vehicle Weight	W_t	3639.80	53.02	0.00	22.91

Table 4.1 – NOMINAL 1958 JAGUAR XK150S MASS PROPERTIES

Obviously if the **sprung weight**, or one of the **unsprung weights**, were to be modified then the **CG** coordinates of the **total vehicle weight** would change; this is a good example of one of those "inescapable" parameter couplings spoken of earlier.

Now that all the equations that relate the parameters to one another have been considered, it is time to deal with the parameters. The parameters of interest in this paper are: **1) Total Weight, 2) Sprung Weight, 3) Unsprung Weight, 4) Forward Unsprung Weight, 5) Rear Unsprung Weight, 6) Total Weight LCG, 7) Sprung Weight LCG, 8) Total Weight VCG, 9) Sprung Weight VCG, 10) Track (average), 11) Front Track, 12) Rear Track, 13) Roll Stiffness (total), 14) Front Roll Stiffness, 15) Rear Roll Stiffness, 16) Roll Axis Height (@ LCG of the Sprung Weight), 17) Front Roll Center Height, and 18) Rear Roll Center Height**[79]. There are other quantities than can be identified as significant parameters in

[78] This "**test configuration**" is based on the "**kerb weight**" (5 Imperial gallons of fuel) as reported in The Autocar of 18 September 1959 (pg 215). To this "**kerb weight**" was "added" a further 9 gallons (Imperial) of fuel (64.8 lb) to bring the vehicle up to the **SAE** standard "**curb weight**" (full fuel). Also" added" were two occupants, a driver and an observer, and a small amount of test equipment (clipboards and stopwatches), totaling a further 327 lb. This brought the vehicle mass properties up to the "**test weight**" condition of about 3540 lb as reported in The Motor of 19 August 1959.

[79] Reference [1], pg.4640. The authors of this reference only identify six primary parameters to the automotive directional stability problem, and then investigate the stability sensitivity to just two: wheelbase and track.

determining automotive directional stability, but other than some tire and aerodynamic characteristics these eighteen quantities are the most basic or **primary** parameters; they are the determining factors for most of those other, **secondary**, parameters. For instance, the roll centers determine the attitude (angle) of the roll axis and, when that is coupled with the **sprung weight LCG/ VCG,** such determines the roll moment arm "$\mathbf{h_r}$" as seen in Fig. 4.1. All of that, plus the roll stiffnesses and the **sprung/unsprung mass properties** as identified in Eq. 4.1, ultimately determines the roll gain "**RG**" which plays such a huge role in the skidpad simulation.

Given this sort of complexity of interaction, each of the parameters to be studied must be carefully considered so as to identify all the ways it can affect the directional stability situation. To give a rather simple example, a **total weight** ($\mathbf{W_t}$) increase of 5% must be reflected as a 5% increase to both the **sprung** ($\mathbf{W_s}$) and **unsprung** ($\mathbf{W_{usf}}$, $\mathbf{W_{usr}}$) **weights**, plus as a 5% increase to the front/rear axle weight loads ($\mathbf{W_f}$, $\mathbf{W_r}$) (see Eq. 4.1 and Eqs. 3.02-3.08); changes in **sprung** and **unsprung weights** mandate a recalculation of the roll gain (actually, a recalculation of the roll equation for which the roll gain is but an inaccurate linearization).

Having discussed the possibility of error resulting from the uncoupling of parameters that have a <u>fundamental</u> and <u>inescapable</u> connection, it is best to add a few words about possible error resulting from the opposite action. In the 1960 paper by Goland and Jindra[80] they created a steady-state mathematical (no mention of a computer simulation or computer usage is made) model of a "**four-wheeled vehicle in a flat turn**". To this model varying input was made of certain parameters: tire width, tire inflation pressure, **longitudinal center of gravity**, front roll stiffness, and rear roll stiffness; all of which was done in order to gauge the corresponding effect on directional stability[81].

Unfortunately when it came to the **longitudinal center of gravity** (**LCG**) Goland and Jindra reasoned as design engineers, not as researchers. They reasoned that as the **LCG** neared one of the axles the spring stiffness (hence roll stiffness) at that axle would have to be increased to deal with the increased **axle weight load**, with the spring stiffness at the other axle correspondingly decreased. Therefore, as the front-to-rear **weight balance** varied they varied the front-to-rear roll stiffness balance in a synchronized lock-step. The resulting effect on the directional stability sensitivity metric reflected, of course, this combined variation of **weight** and roll stiffness variation, and was not as useful as would have been the case had those parameters been appropriately decoupled.

---

[80] Reference [13]. Essentially the basic concept of this reference paper was the same as this author's paper, except lacking the emphasis on **mass properties**. The authors of this paper consider the directional stability effect of a variation in the vehicle **LCG**, but confused the issue by including a simultaneous variation in suspension stiffness. However, they also experimented with considerable variation in tire characteristics, which is one reason why this author neglected a similar tire investigation in this paper.

[81] Directional stability for the Goland and Jindra paper was measured by a metric termed the "**Control Parameter**" (symbolized as "**K**", with units of "$\mathbf{sec^2/ft^2}$").

5 – SENSITIVITY ANALYSIS

There are a number of definitions of "**sensitivity analysis**"; one of the more common is that "**sensitivity analysis is…how the uncertainty in the output of a mathematical model…can be…allocated to…uncertainty in its inputs**"[82]. In other words, sensitivity analysis is the determination of how the math model output varies in response to variation in each of its input parameters. Such knowledge of a model's sensitivity to input variation can be of use in determining the "**robustness**" of the model, or in determining where to focus attention in the design process in order to most efficiently obtain desired results.

There are a number of sensitivity analysis methods, of which the more sophisticated are beyond this author's ability to understand, let alone apply. However, there are at least three methods which are well within this author's purview: **Partial Differentiation Analysis**, **OAT Analysis**, and **Input-Output Regression Analysis**. **Partial Differentiation Analysis** requires that the math model consists of a single explicit equation which may be subjected to a partial differentiation with respect to each of the input variables[83]; since the matter of automotive directional stability is not readily expressible as a single explicit equation this type of sensitivity analysis method will be of no further concern in this paper.

That leaves the **OAT** and **Input-Output Regression** methods for consideration. The **OAT**, or "**One-At-a-Time**", is perhaps the simplest and most common method of sensitivity analysis. Each input variable of interest is varied individually while holding all other input variables constant at their nominal values. The variation in output variable per variation in each input variable is noted, and establishes the "**sensitivity**" of the output to that input variable. However, if not somehow accounted for, the presence of certain inescapable interactions between input variables, plus the existence of highly non-linear relationships, can render **OAT** sensitivity analysis results misleading[84]. This was discussed in the present automotive context earlier in Chapter 4, but could bear some emphasis…

As a simple example of how the presence of complex interactions between input variables can render an **OAT** analysis result erroneous, given the automotive context of this paper, consider the matter of the effect on directional stability of variation of the **sprung weight**. If the **sprung weight** ("$\mathbf{W_s}$") were to increase then both the **total vehicle weight** ("$\mathbf{W_t}$") and **total vehicle VCG** ("$\mathbf{h_{cg}}$") must increase, along with the **front** ("$\mathbf{W_f}$") and **rear axle** ("$\mathbf{W_r}$") **weight loads**. Only slightly less obvious is the fact that the roll gain equation must change, resulting in a proportionately greater roll gain ("**RG**") value. All of this results in a considerably greater roll angle ("$\boldsymbol{\theta_s}$") and "**weight transfer**" per **g** of lateral acceleration, thereby substantially affecting the directional stability as measured by the Understeer Gradient "$\mathbf{K_{us}}$". While it is a great advantage of computer simulation models that various parameters can be "**decoupled**" so

---

[82] https://en. Wikipedia.org, "Sensitivity Analysis", 6/07/2021.
[83] Ibid.
[84] Ibid.

that parametric variation effects can be explored individually, if certain unavoidable interactions between variables, such as the interactions between **sprung weight** and "$\mathbf{h_{cg}}$" or "**RG**" are ignored, then the resulting **OAT** Sensitivity Analysis gradient ($\mathbf{\Delta K_{us}/\Delta W_s}$) will be worthless.

In an attempt to avoid such an occurrence for the **OAT** Sensitivity Analysis conducted for this paper, each input parameter studied will be accompanied by a search for unavoidable interactions with other parameters through a consideration of the relationships expressed in Eq. 3.02 to 3.08, Eq. 4.1, the geometric relationships expressed in Fig. 4.1 ("Side" view), the roll stiffness relations expressed in Eq. 4.2-4.3, and the **mass property** relations expressed in Eq. 4.4-4.7. When an unavoidable effect on another parameter is identified, then that effect will be accounted for through appropriate variation of the affected parameter.

That leaves the matter of **Input-Output Regression Sensitivity Analysis**, which may serve as a useful adjunct to the **OAT** in addressing the concern that the non-linear nature of automotive dynamics relationships (see Fig. 3.6) could render any resulting **OAT** analysis linear gradient ($\mathbf{\Delta K_{us}/\Delta W_t}$, etc.) inaccurate and useless. **Input-Output Regression Analysis** consists of subjecting the various resulting parameter "**input/output**" value sets to a linear regression (i.e., fitting a straight line to the data plot). If the **Coefficient of Determination** ($\mathbf{R^2}$) of the fit is sufficiently close to "**1**" then the linearity of the relationship is such that the gradient (slope of the fitted line) may be considered accurate enough to be useful.

There may be other possible deficiencies regarding the sensitivity analysis as performed for this paper. Due to limitations of time and resources, a sensitivity analysis variation in input variable (parameter) is often of a very narrow or "**local**" range, and therefore may not produce results of "**global**" applicability; since the automotive subject of this paper (road going passenger cars) does not have the range of variation in size typical of aircraft or ships the general range of input parameter variation used in this paper (**±10%**) should be sufficient.

Finally there are problems arising from practical matters, as opposed to theoretical. Limitations in time and resources may impact the amount of data that can be processed; the length of time it takes to make a model "**run**" may preclude processing the amount of data necessary to acquire a high degree of confidence in the results. Other than the acquisition of more time and resources, there is little that can be done to negate such practical concerns; it is the lack of time and resources which will have the greatest negative effect on this paper's results.

6 – CONCLUSIONS

The following "**Table 6.1 – DIRECTIONAL STABILITY SENSITIVITY to 18 PARAMETERS**" presents the essence of this paper's results. Each of the 18 specified parameters was varied **OAT** by percent in discrete steps over the range "**-10%, -5%, 0%, +5%, +10%**" with the consequent variation in the Understeer Gradient ("$\mathbf{K_{us}}$") noted. From this information a dimensionless "**Directional Stability Sensitivity Metric**" was determined as a "**Percent Understeer Gradient Variation per Percent Parameter Variation**" (**%$\mathbf{K_{us}}$/%Parameter**). Due to the inherently nonlinear nature of the underlying relationships the sensitivity of the directional stability to each parameter could not be accurately represented by a single value; instead a discrete value had to be determined for each of the "**-10%, -5%, 0%, +5%, +10%**" steps in parameter variation. That is to say, there are 5 Directional Stability Sensitivity Metric values for each parameter investigated. **Input-Output Regression Analysis** was employed to quantify the degree of non-linearity for each of the 18 parameters; that is, the percent parameter variation versus the corresponding stability metric ("$\mathbf{K_{us}}$") was plotted and a straight line "**fitted**" to the data by means of regression analysis, with the corresponding "**Coefficient of Determination**" ("$\mathbf{R^2}$") quantifying the linearity.

The results are generally in accord with what could have reasonably been expected beforehand. The top 4 most influential parameters are **mass properties**, with **mass properties** taking the 1, 2, 3, 4, 8, 10, 11, 13, and 16 spots. The **total vehicle weight LCG** ("$\mathbf{l_f}$") is by far the most influential parameter, followed by the **sprung weight LCG** ("$\mathbf{l_{fs}}$"). The non-linearity of the underlying relations is well illustrated by these **LCG** parameters as moving the **CG** aft (+5%, +10%) has significantly greater adverse effect (increased instability) than moving the **CG** forward (-5%, -10%) has beneficial effect (increased stability). The most influential **non-mass property** parameters (taking the 5 and 6 spots) are the vehicle track ("**t**", both forward and rear track varying in sync) followed by the forward vehicle track ("$\mathbf{t_f}$", varying discretely), which tend to only be a 4th to a 10th as influential as the **LCG** parameters.

Notable by its omission from the 18 parameters investigated is the wheelbase ("$\mathbf{l_{wb}}$"), and it is appropriate that some explanation be given as to why this is so. Given the nature and simplifications inherent in the skidpad simulation the principal effects of wheelbase variation are simply not "picked up". A major effect of wheelbase variation is a variation in the yaw inertia, but as the skidpad simulation is of the steady-state condition there is no angular acceleration about the yaw axis, rendering any variation in the yaw inertia moot. Another significant effect of wheelbase variation is on the resolution of the tire traction force into lateral and longitudinal components (see "**Appendix G – Angularity, Resolution of Tire Traction Force(s) into Components**"). Since the skidpad radius of 100 feet renders the lateral component a relatively huge portion of the traction force, regardless of any reasonable variation in vehicle wheelbase, the angularity was neglected with the resultant force being taken as equal to the lateral traction force. This simplification had negligible effect on all the other parameter investigations while significantly simplifying and speeding up the construction of the skidpad simulation.

The neglect of the angularity of the traction force was not the only skidpad simplification, which included the omission of camber forces, aligning moments, and aerodynamic forces; all of that may be significant in determining exact lateral performance levels for specific vehicles, but should not significantly affect this general ranking of parameters. Although the skidpad program can continue to generate data past the maximum lateral acceleration level(s) (see "**Appendix H – Maximum Lateral Acceleration vs. Parameter Variation**") of the subject vehicle, such data was regarded as extraneous and not included in the analysis. The consequent directional stability sensitivity results are summarized in the following Table 6.1….

				Directional Stability Metric % Variation per % Variation in Parameter About the Norm of 0%:					Variation Linearity:	
Sensitivity Ranking	Parameter	Symbol	Units	-10%	-5%	0%	+5%	+10%	R^2	Comments:
3	Total Vehicle Weight	W_t	lb	6.959270	8.857013	0.0	-17.022694	-19.499691	0.9617	Add Wt much more adverse than reducing Wt is beneficial (nonlinear)
4	Sprung Weight	W_s	lb	4.025712	4.097616	0.0	-14.066874	-16.478707	0.8942	Same, but slightly less influential (very nonlinear)
13	Unsprung Weight	W_{us}	lb	1.129433	1.164876	0.0	-1.488951	-1.618627	0.9939	Same, but much less influential
16	Fwd Unsprung Weight	W_{usf}	lb	-0.832645	-0.835524	0.0	0.860770	0.860069	0.9999	Add Fwd Unsprung Wt benefits directional stability (almost linear)
11	Rear Unsprung Weight	W_{usr}	lb	1.951758	1.995188	0.0	-2.298037	-2.427083	0.9961	Reducing Rear Unsprung Wt benefits direcional stability
1	Total Vehicle Wt LCG	l_f	in	21.229745	21.380853	0.0	-26.154503	-27.961410	0.9939	Moving Total Vehicle LCG forward greatly benefits stability
2	Sprung Weight LCG	l_{fs}	in	17.775241	17.901333	0.0	-21.898642	-23.411529	0.9939	Same, but slightly less influential
8	Total Vehicle Wt VCG	h_{cg}	in	1.718460	1.971002	0.0	-4.449144	-4.733857	0.9388	Reducing Total Vehicle VCG benefits stability
10	Sprung Weight VCG	h_{cgs}	in	1.708863	1.929629	0.0	-3.732691	-4.084553	0.9480	Reducing Sprung Wt VCG slightly less beneficial to stability
5	Track	t	in	-7.781033	-6.415679	0.0	2.937690	2.579838	0.9869	Increasing Track (Front & Rear) benefits stability
6	Track Forward	t_f	in	-5.420090	-4.594291	0.0	2.907376	2.516702	0.9558	Increasing Track Forward slightly less beneficial to stability
17	Track Rear	t_r	in	-1.059392	-0.654726	0.0	0.442563	0.325437	0.9278	Increasing Track Rear much less beneficial to stability
14	Roll Axis Height	h_{ra} @ l_{fs}	in	0.998619	0.981537	0.0	-0.983272	-1.078002	0.9988	Decreasing Roll Axis Height beneficial to stability
15	Roll Center Forward	h_{rcf}	in	-0.993205	-0.900353	0.0	0.835581	0.823028	0.9963	Nom Fwd RC neg (below ground); "increasing" means make less neg
12	Roll Center Rear	h_{rcr}	in	1.820285	1.816004	0.0	-1.935244	-2.130152	0.9972	Nom Rr RC is pos (abv ground, more common), decreasing beneficial
18	Total Vehicle Roll Stiff	K_{roll}	ft-lb/deg	-0.701431	-0.564527	0.0	0.402838	0.398468	0.9711	Increasing Total Vehicle Roll Stiffness benefits directional stability
7	Roll Stiffness Forward	K_{froll}	ft-lb/deg	-4.894215	-4.255439	0.0	3.116554	2.899405	0.9775	Increasing Roll Stiffness Forward benefits directional stability
9	Roll Stiffness Rear	K_{rroll}	ft-lb/deg	2.956143	2.930092	0.0	-3.264498	-3.452787	0.9979	Decreasing Roll Stiffness Rear benefits directional stability
				Pos Stability Metric means improved stability (green color)						
				Neg Stability Metric means reduced stability (rose color)						

Table 6.1 – DIRECTIONAL STABILITY SENSITIVITY to 18 PARAMETERS

As noted this analysis summary does not include the wheelbase parameter due to skidpad modeling simplifications which delete most of the wheelbase effects. What was left for a wheelbase variation to affect was the roll moment arm as calculated per the geometry of Figure 4.1; over the range of variation of ±10% this variation proved to be around a ±0.1 inch. Not only did this prove to have an insignificant effect on the directional stability metric, but the effect was the reverse of reasonable expectation. Regarding that reasonable expectation, researchers Anbazhagan, Satheesh, and Gopalakrishnan "proved" in their paper that the ratio of track stability effect to wheelbase stability effect was about 1.65:1[85]. Based on this information, an "educated guess" would have the wheelbase ranking somewhere around 10th place for being influential on directional stability:

Ranking	Parameter	Symbol	Units	-10%	-5%	0%	5%	10%
10	Wheelbase	l_{wb}	in	-4.714470	-3.887071	0.0	1.779354	1.562562

Table 6.2 – ESTIMATED WHEELBASE DIRECTIONAL STABILITY SENSITIVITY

As noted previously, the variation in parameters also resulted in a variation in maxium lateral acceleration levels ("$\mathbf{a_{ymax}}$"), which determined the point of skidpad simulation "termination". Although the skidpad simulation is capable of generating data beyond this point, for this directional stability study any such data was discarded. This action would seem to be common sense, but further justification was given by Goland and Jindra: "**...onset of gross skidding...represents a limiting condition beyond which...analysis loses validity...**"[86].

The Goland and Jindra paper comes closest to matching this paper in intent of any relevant paper found by this author, with the possible exception of the paper by Anbazhagan, Satheesh, and Golpalakrishnan (Reference [1]). The Goland and Jindra paper makes many of the same simplifications, and provides a justification for them. Their approach, like this author's, does not include aligning torque: "**For the purposes of this study, these torques do not exert a significant influence.**"[87]. They also ignore, as does this paper, the effect of minor disturbance forces that result in small pitch and roll motions: "**This is permissible on the grounds that these motions exert only a minor influence on the directional stability**..."[88]. They also neglect roll steer and roll camber ("...**to further reduce the number of variables in the calculations**...")[89], although they do introduce a crude general method of accounting for camber thrust based on a 1957 tire study by researchers Smiley and Horne[90]. They say that they regard the steering wheel as fixed and the steering linkage as rigid[91], but in practice they actually vary the steering angles per 4 equations (4-wheel steering?) prominently displayed[92].

---

[85] Reference [1], pg. 1639.
[86] Reference [13], pg. 4.
[87] Ibid.
[88] Ibid, pg. 3.
[89] Ibid, pg. 11.
[90] Ibid, pg. 5. Smiley, R.F.; and W.B. Horne, "**Mechanical Properties of Pneumatic Tires with Special Reference to Modern Aircraft Tires**", **NACA TN** 4110, 1957. Goland and Jindra make much use of this paper for their tire model, using general equations not representative of any specific tire (maybe not even an automotive tire!) for their tire model.
[91] Reference [13], pg. 2.
[92] Ibid, pg. 7. It may be that Goland and Jindra are allowing for the possibility of 4-wheel steering, or perhaps rear wheel steering, although that would seem to be very advanced thinking for c. 1960).

This paper also varies the steering angles, but recognizes the existence of such only at the two front wheels, in accord with the specific "**High Speed Ackermann Steering Geometry**" equations (see Fig. 3.5 for the general formulation) of:

$$\boldsymbol{\delta_o = 57.3\,\frac{l_{wb}}{R_o} + \psi_{fo} - \psi_{ro}} \quad \text{(EQ. 6.1)}$$

$$\boldsymbol{\delta_i = 57.3\,\frac{l_{wb}}{R_i} + \psi_{fi} - \psi_{ri}} \quad \text{(EQ. 6.2)}$$

This would indicate that the steering wheel in not "fixed" in this paper (it wasn't even in the Goland and Jindra paper, despite what they said), but "free" to rotate to whatever position is required to maintain course along the skidpad circumference; this would seem to reflect reality as steering corrections would be necessary for an actual car to maintain course as lateral acceleration levels (velocity) increases.

Although the Goland and Jindra paper is primarily concerned with steady-state, a.k.a. "**static**", stability, they also are able to link this stability condition to that transient stability condition known as "**dynamic**" or "**oscillatory**" (per Walter Bergman) through their "**control parameter**" (symbolized as "**K**"). That "**K**" parameter is "plugged" into a relationship involving velocity ("**V**") and at least two other parameters ("**ч**", "**η**"), which might be easy to decipher if only a symbolism list had been provided. If further study of the Goland and Jindra paper does lead to clarification of their transient stability methodology, then it may point the way for an expansion of the content of this paper at some future date.

In any case, they investigate the effect of variation of four parameters upon steady-state stability: tire tread width, tire inflation pressure, vehicle **LCG**, and suspension spring stiffness. However, this paper does not investigate the effect of any tire parameter variation, and Goland-Jindra couple their variation of **LCG** to a simultaneous spring stiffness variation, so there is no result directly comparable to the results of this paper. It is the construction of their steady-state model, as a 3 **DOF** (Yaw, Slip. Roll) of incremental velocity variation on a 150 ft radius turn that takes into account the effect of varying normal loads on tire cornering stiffness, that correlates well with this paper's methodology, except they do state that their modeling is of the linear variety[93].

A major credit of this paper's methodology is that it is nonlinear, and is valid well beyond 0.3 g's right on up to the maximum lateral acceleration. The vehicle model is of the 3 DOF (side slip, yaw, and roll) two mass, sprung and unsprung, type. The roll motion is about a realistic roll axis, resulting in accurate normal loads and correct tire slip angle and cornering stiffness values. The simplifications made do not significantly affect the results of this paper, and the skidpad simulation is such that it can readily be subject to future enhancement by the addition of camber, alignment torque, and aero (lift effect on normal loads) effects.

[93] Reference [13], pg. 2.

The paper by Anbazhagan, Satheesh, and Gopalakrishnan is even more difficult to compare with this paper, whether it is with regard to results or methodology. They state they have constructed a program in the "**C**" language for their stability analysis[94], but they say nothing of what simplifications may have been involved, whether the modeling was linear or non-linear, etc. They do not supply a listing of the program, or samples of their input/output files. They identify 6 parameters as relevant to the directional stability problem (wheelbase, track, **CG**, lateral stiffness, vertical damping, and wheel radius)[95], but they only supply results of their stability analysis for 2 parameters, wheelbase and track[96], in the form of plot(s):

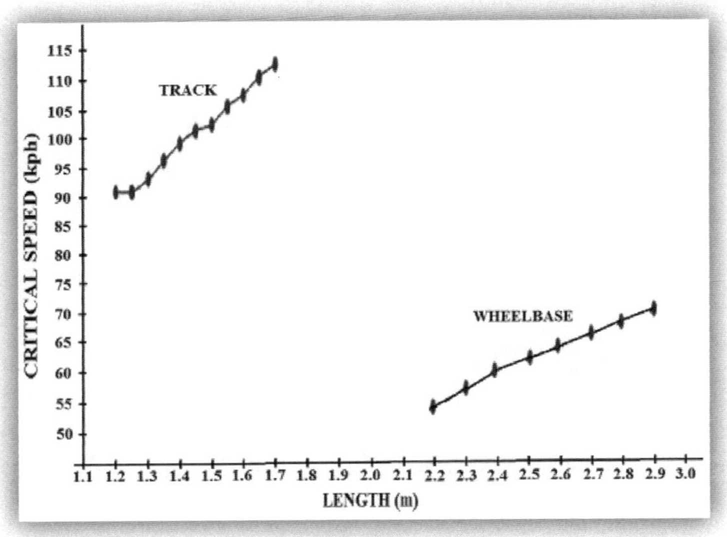

Figure 6.1 - ANBAZHAGAN, SATHEESH, and GOPALAKRISHNAN STABILITY RESULTS[97]

Note that the stability metric utilized by Anbazhagan, Satheesh, and Gopalakrishnan is "**Critical Speed**", which is not directly comparable with "**Understeer Gradient**" as used by this paper. Ultimately it can be said that the paper by Anbazhagan, Satheesh, and Gopalakrishnan is of very limited utility in "validation" of the methodology or results of this paper.

With regard to the results of this paper, which can be condensed to just the directional stability-sensitivity-to-parameter dimensionless (%Chg-Stability/%Chg-Parameter) metrics of Table 6.1, those results can be useful in making decisions in preliminary design. This statement is echoed by Goland and Jindra in their paper: "...**results will be of value for preliminary design**..."[98]. This is the ultimate purpose of this paper, although it is hoped that a lot of interesting information was provided along the way to serving that purpose.

---

[94] Reference [1], pg.4639.
[95] Reference [1], pg. 4640.
[96] Ibid, pg. 4639.
[97] This was presented as two different plots, but those two plots were combined into one by this author.
[98] Reference [13], pg. 1.

REFERENCES

[1] Anbazhagan, R.; B. Satheesh, and K. Gopalakrishnan; "Mathematical Modeling and Simulation of Modern Cars in the Role of Stability Analysis", Indian Journal of Science and Technology, May 2013.

[2] Bastow, Donald; Car Suspension and Handling, Pentech Press Ltd; London, UK, 1980.

[3] Bergman, Walter; "The Basic Nature of Vehicle Understeer-Oversteer", SAE Paper 650085; Warrendale, PA, 1965.

[4] Bundorf, R. Thomas; and R.L. Leffert; "Cornering Compliance Concept for Description of Vehicle Directional Control Properties", SAE Paper 760713; Warrendale, PA, 1976.

[5] Bundorf, R. Thomas; "The Influence of Vehicle Design Parameters on Characteristic Speed and Understeer", SAE Paper 670078; Warrendale, PA, 1967.

[6] Bundorf, R. Thomas; "Vehicle Dynamics at General Motors Corporation 1952-1980", SAE Automotive Dynamics and Stability Conference (Presentation); Troy, MI, 2000.

[7] Campbell, Colin; Design of Racing Sports Cars, Robert Bentley Inc.; Cambridge, MA, 1976.

[8] Campbell, Colin; The Sports Car, Robert Bentley Inc.; Cambridge, MA, 1969.

[9] Campbell, Colin; New Directions in Suspension Design, Robert Bentley Inc.; Cambridge, MA, 1981.

[10] Costin, Michael; and David Phipps; Racing and Sports Car Chassis Design, Robert Bentley Inc.; Cambridge, MA, 1967.

[11] Gent, Alan N.; and J.D. Walter, The Pneumatic Tire, US DOT/NHTSA, DOT HS 810 561; Washington, DC; 2006.

[12] Gillespie, Thomas D.; Fundamentals of Vehicle Dynamics, SAE R-114; Warrendale, PA, 1992.

[13] Goland, M.; and F. Jindra, "Directional Stability and Control of a Four-Wheeled Vehicle in a Flat Turn", SAE Paper 600208; Warrendale, PA, 1960.

[14] Hucho, Wolf-Heinrich; Aerodynamics of Road Vehicles, SAE R-177, Warrendale, PA, 1988.

[15] Karogal, Indrasen; E. Harry Law, and Bashah Ayalew; "An Iterative Approach for Steady State Handling Analysis of Vehicles", Proceedings of the ASME 2008 IDETC/CIE Conference, DETC2008-50033; Brooklyn, NY, 3-6 August 2008.

[16] Kasprzak, Edward M.; and David Gentz; "The Formula SAE Tire Test Consortium – Tire Testing and Data Handling", SAE Paper No. 2006-01-3606; Warrendale PA, 2006.

[17] Lamar, Paul; "Aerodynamics & the Group Seven Racing Car", The Aerodynamics of Sports & Competition Automobiles, Proceedings of AIAA Symposium; Los Angeles, CA, 20 April 1968.

[18] Lamar, Paul; "More About Cornering Power", Road & Track, October 1969; pp. 127-130.

[19] Milliken, William F.; and Douglas L. Milliken; Race Car Vehicle Dynamics, SAE R-146; Warrendale, PA, 1995.

[20] Milliken, William F.; and Douglas L. Milliken; Chassis Design Principles and Analysis, SAE R-206; Warrendale, PA, 2002.

[21] Nalecz, Andrzej G.; "Investigation into the Effects of Suspension Design on Stability of Light Vehicles", SAE Paper 870497; Warrendale, PA, 1987.

[22] Nalecz, Andrzej G.; "Development and Validation of Light Vehicle Dynamics Simulation (LVDS)", SAE Paper 920056; Warrendale, PA, 1992.

[23] Reimpell, Jörnsen; Helmut Stoll, and Jürgen W. Betzler; The Automotive Chassis; SAE R-300, Butterworth-Heinemann; Jordan Hill, UK, 2001.

[24] Segel, Leonard; "Theoretical Prediction and Experimental Substantiation of the Response of the Automobile to Steering Control", Institute of Mechanical Engineers; London, UK, 1956.

[25] Segel, Leonard; "Research in the Fundamentals of Automobile Control and Stability", SAE Paper 570044; Warrendale, PA, 1957.

[26] Speckhart, Frank H.; "A Computer Simulation for Three-Dimensional Vehicle Dynamics", SAE Paper 730526; Warrendale, PA, 1973.

[27] Steeds, W.; Mechanics of Road Vehicles, Iliffe & Sons Ltd; London, UK, 1960.

[28] Taborek, Jaroslav J.; Mechanics of Vehicles, Penton Publishing Co.; Cleveland, OH, 1957.

[29] Tamaddoni, Seyed Housein; "Optimal Vehicle Stability Control with Driver Input and Bounded Uncertainties", Virginia Polytechnic & State University; Blacksburg, VA, 2011.

[30] Wiegand, Brian Paul; "The Basic Algorithms of Mass Properties Analysis and Control", SAWE Paper #2064; Hartford, CT, 1992.

[31] Wiegand, Brian Paul; "Automotive Mass Properties Estimation", SAWE Paper #3490; Virginia Beach, VA, 2010.

[32] Wiegand, Brian Paul; "Mass Properties and Automotive Lateral Acceleration", SAWE Paper #3528; Houston, TX, 2011.

[33] Wiegand, Brian Paul; "Automotive Lateral Dynamics and Mass Properties", SAWE Training Seminar, SAWE Regional Conference (Irving, TX; 2017) and SAWE International Conference (Norfolk, VA; 2019).

[34] Wong, Jo Yung; Theory of Ground Vehicles, John Wiley & Sons Inc.; Hoboken, NJ, 2008.

[35] ____________; "Surface Vehicle Recommended Practice: Vehicle Dynamics Terminology", SAE J670_200801; Warrendale, PA, 2008.

[36] _____________; SAWE Weight Engineers Handbook, Society of Allied Weight Engineers; Los Angeles, CA, Rev. May 2002.

[37] Milliken, William F.; Equations of Motion, Bentley Publishers, Cambridge, MA; 2006.

SAWE Paper No. 3765
Category Number 31.0

AUTHOR'S BIOGRAPHICAL SKETCH

Brian Paul Wiegand, now retired, was a Senior Weight Engineer and Mass Properties Handling Specialist for the Mass Properties Analysis and Control Group of Northrop Grumman Corporation, Bethpage, NY. He is a 1972 graduate of Pratt Institute, Brooklyn, NY, and a licensed Professional Engineer registered in the State of New York (#58470). He continues to be an active member of the Society of Allied Weight Engineers and of the Society of Automotive Engineers, and in 2009 he reaffirmed his long-standing interest in automotive dynamics by attending the **SAE** Seminar "**Vehicle Dynamics for Passenger Cars and Light Trucks**" (Troy, MI; August 11-13). He has presented eight **SAWE** papers, two award winning:

"**Mass Properties and Automotive Longitudinal Acceleration**" (**SAWE** #1634, 1984)

"**The Basic Algorithms of Mass Properties Analysis & Control**" (**SAWE** #2067, 1992)

"**Automotive Mass Properties Estimation**" (**SAWE** #3490, "**Special Merit Award**", 2010)

"**Mass Properties and Automotive Vertical Acceleration**" (**SAWE** #3521, "**Hackney Award**", 2011)

"**Mass Properties and Automotive Lateral Acceleration**" (**SAWE** #3528, 2011)

"**Mass Properties and Advanced Automotive Design**" (**SAWE** #3602, 2015)

"**Mass Properties and Automotive Crash Survival**" (**SAWE** #3634, 2015)

"**Colin Chapman and Automotive Mass Properties**" (**SAWE** #3640, 2015)

He has also published six **SAWE Journal** ("**Weight Engineering**") articles, including two award winning:

"**The Weight and CG Implications of Obtaining Maximum Automotive Lateral Acceleration Levels**" (Winter 1982-83)

"**The Dynosphere: 1932**" (Spring 2011)

"**The Mystery of Automotive POI Values**" (Winter 2011)

"**The Gyrauto: 1935**" (Spring 2012)

"**Ancient Mass Properties Engineering**" (Winter 2011-12, "**Best Journal Article 2011-12**")

"**An Uncertainty Regarding CG Uncertainty**" (Winter 2012-13, "**Best Journal Article 2012-13**")

He also has presented one **SAE** paper: "**Estimation of the Rolling Resistance of Tires**" (**SAE** #2016-01-0445, 2016), and one **ASNE** book review: "**Big Gun Monitors by Ian Buxton**" (**Naval Engineers Journal**, December 1981). He has twice served as instructor for the **SAWE** Training Seminar "**Automotive Lateral Dynamics and Mass Properties**", given initially at the 2017 **SAWE** Regional Conference (Irving, TX), and then again at the 2019 **SAWE** International Conference (Norfolk, VA). Lastly, he is the recipient of the **2011 President's Award** for contributions to the **SAWE,** and of two Northrop Grumman "**TAP's**" (**Timely Awards Plan**, 1996 and 2006).

APPENDICES

APPENDIX A – SYMBOLISM

A Coefficient in a quadratic equation, like that for roll angle (θ_s) of sprung mass undergoing lateral acceleration (a_y): $\theta_s = A\, a_y^2 + B\, a_y + C$.

a Usually stands for acceleration (ft/sec^2, in/sec^2, m/sec^2, g's) as per Newton's Second Law: "**F = ma**", but can also symbolize the distance between the front axle and the vehicle **CG** as per **SAE** J670_2008 (para. 4.1.22) (in, cm).

α Greek lower case letter "**alpha**", usually stands for an angular acceleration such as in the rotational version of Newton's Second Law: "**T = Iα**" (deg/sec^2, rad/sec^2). In the transient case a turning automobile rotates about a vertical axis of varying location with a varying angular velocity. As the transient case evolves into the steady-state case the angular acceleration dies out leaving an angular velocity about a vertical axis through the **CG** (deg/sec^2).

ABS "**Anti-skid Braking System**" refers to automatic systems for brake modulation intended to prevent wheel "**lock-up**" under hard and/or low traction braking conditions.

ADAMS "**Automated Dynamic Analysis of Mechanical Systems**" is a popular multi-body dynamics simulation software package. **ADAMS** was originally developed by Mechanical Dynamics Inc., which has been acquired by the **MSC Software Corporation**. **ADAMS** is modular, allowing linkage with other software packages and other customization to meet specific industry requirements. Some existing "semi-customized" modules are **ADAMS**/Car, **ADAMS**/Engine, **ADAMS** /Aircraft, and **ADAMS/**Machinery.

ADAMS/Car An **ADAMS** module developed specifically for the dynamic analysis of automotive systems.

AF Stands for "**Air Force**", which in the context of this paper is short for the "**United States Air Force**" (**USAF**).

A_f The vehicle frontal area (ft^2, m^2).

AIAA The "**American Institute of Aeronautics and Astronautics**" is the professional society for the field of aerospace engineering. The **AIAA** was founded in 1963 from the merger of two earlier societies, and today has about 35,000 members.

ANSYS Ansys Inc. is an American company based in Canonsburg, PA. The **ANSYS** software is a Finite Element Modeling (**FEM**) based program for the analysis and development of structures, fluid systems, semiconductors, electromagnetic systems, and optical systems. The company was founded 50 years ago by John Swanson, who sold out his interest in 1993; the company went public in 1996.

A_p The vehicle plan area (ft^2, m^2). Note that it is common practice in the automotive engineering world to finesse the coefficient of lift "**C_l**" so that it can be used in conjunction with the frontal area and still generate correct lift force values.

AR A tire's "**Aspect Ratio**", which is the ratio of the tire shoulder height to the tire section width times 100: "**$AR = S_H/S_N \times 100$**" (dimensionless).

ARB Stands for "**Anti-Roll Bar**", which is a suspension component that increases the roll stiffness while causing relatively little (or no) increase in vertical spring rate. The first **ARB** patent was awarded to S.L.C. Coleman of Canada on April 22, 1919.

ARP "**Active Rollover Protection**" refers to a system that detects a condition of impending rollover and selectively applies the vehicle brakes so as to create a counteracting roll moment. Future **ARP** systems may create a counteracting roll through the action of a fully active suspension.

Auto-CAD A "**Computer Aided Design**" (**CAD**) and "**Computer Aided Engineering**" (**CAE**) software package marketed by AutoDesk Inc. of San Rafael, CA.

AWD "**All wheel drive**", generally meaning a constant drive configuration of **4WD** (no **RWD** option).

a_y An acceleration in the lateral direction (ft/sec^2, m/sec^2), usually in gravity units (g's).

ay_{fpot} The potential lateral acceleration level attainable at the front axle based on the potential lateral traction forces available at the front tires (g's).

ay_{rpot} The potential lateral acceleration level attainable at the rear axle based on the potential lateral traction forces available at the rear tires (g's).

a_{ymax} The maximum lateral acceleration attainable before a vehicle slides off a skidpad, either nose first (**understeering**) or tail first (**oversteering**) (g's).

$a_{yoverturn}$ The lateral acceleration level at which a vehicle would overturn, which is not usually attainable as most vehicles will reach "**a_{ymax}**" first (g's).

B Coefficient in a quadratic equation, like that for roll angle (**θ_s**) of sprung mass undergoing lateral acceleration due to a lateral force (**F_y**): **$\theta_s = A\,F_y^2 + B\,F_y + C$.**

b The basic tire coefficient of traction potential (dimensionless) as used in the "Lamar" equation for a tire's lateral force potential: "**$F_{ypot} = (b - mN)N$**" (dimensionless). Also can stand for the distance from the vehicle **CG** to the rear axle line per the **SAE J670_2008** (para. 4.1.22) (in, cm).

β The Greek lower case letter "**beta**" meaning the "**attitude angle**" at the vehicle **CG** between the longitudinal axis and the velocity vector (degrees).

BASIC "**Beginner's All-purpose Symbolic Instruction Code**" is a family of general purpose programming languages whose design emphasizes ease of use. The original **BASIC** was created by John G. Kenny and Thomas E. Kurtz at Dartmouth College in 1964 with the intent to enable students in fields other than science, engineering, and mathematics to utilize computers.

b_{if} The basic or peak lateral coefficient of traction at the inner front tire (dimensionless).

b_{ir} The basic or peak lateral coefficient of traction at the inner rear tire (dimensionless).

BNP "**Bakker-Nyborg-Pajecka**" or "**Magic Formula**" is the predominant model of tire behavior; variations of the "**Magic Formula**" are used professionally by competitive racing organizations, tire manufacturers, and automotive corporations the world over.

b_{of} The basic or peak lateral coefficient of traction at the outer front tire (dimensionless).

b_{or} The basic or peak lateral coefficient of traction at the outer rear tire (dimensionless).

C Coefficient in a quadratic equation, like that for roll angle (θ_s) of sprung mass undergoing a lateral acceleration force (F_y): $\theta_s = A\,F_y^2 + B\,F_y + C$.

CAD "**Computer Aided Design**".

CAE "**Computer Aided Engineering**".

CAL The **CAL** (**Cornell Aeronautical Laboratory**) got started as private defense contractor Curtiss-Wright's aero laboratory in WW II. Post-war, Curtiss-Wright donated the lab to Cornell University to operate "***as a public trust***". The lab operated under the name **Cornell Aeronautical Laboratory** from 1946 until 1972. During the late 1960's/early 1970's universities came under criticism for conducting war-related research, and in 1972 Cornell reorganized the lab as the "***for-profit***" **Calspan Corporation** and sold all its stock to the public.

CAM "**Computer Aided Manufacturing**".

CarSim A commercial software package for the simulation of automotive performance produced and distributed by Mechanical Simulation Company using technology that originated at the University of Michigan Transportation Research Institute (**UMTRI**) in Ann Arbor, MI.

CATIA The oldest and most respected software of its kind, **CATIA** started in 1977 as an in-house development of the French aircraft company ***Avions Marcel Dassault***. Today it is one of the most widely used **CAD/CAE/CAM** software products in the world.

C_c The "**Cornering Coefficient**" is the "normalized" (divided by the load) "**Cornering Stiffness**": "**C_s/N**" (lb/deg/lb).

C/D Car and Driver magazine is a major **US** reviewer and tester of new production automobiles; it was founded as "Sports Car Illustrated" in 1955.

C_D The Stuttgart Institute of Technology formula "dynamic" coefficient of tire rolling resistance (lb/lb-mph$^{2.5}$).

C_d The aerodynamic coefficient of drag (dimensionless).

CG The "**center of gravity**" of a vehicle or other object.

C_l The aerodynamic coefficient of lift (dimensionless).

CP The aerodynamic "**center of pressure**" of a vehicle or other object.

CSMP The "**Continuous System Modeling Program**" was marketed in 1976 by inventor Frank H. Speckhart for the simulation of complex systems defined by sets of differential equations solvable by numerical integration.

C_S The Stuttgart Institute of Technology formula "static" coefficient of tire rolling resistance (lb/lb).

C_s The "**Cornering Stiffness**" (a.k.a. "Cornering Power", a misnomer), is the partial derivative "$\partial F_y/\partial \psi$" of the tire function "$F_y = f(\psi)$" at a constant normal load "**N**" (lb/deg, lb/rad, etc.).

C_{sf} The cornering stiffness value at the front axle (the sum of the stiffness values of the front tires) (lb/deg, N/rad).

C_{sr} The cornering stiffness value at the rear axle (the sum of the stiffness values of the rear tires) (lb/deg, N/rad).

D The aerodynamic drag force (lb, kg).

DOF "**Degrees of Freedom**", within the context of this paper, refers to the number of ways a physical body can move; for a simple homogeneous **mass** there are 6-**DOF**: three directions and three rotations.

DOT The "**Department of Transportation**" is a Cabinet level department of the United States government established by act of Congress on 15 October 1966. It has the responsibility for the regulation of all matters relating to transportation, and consists of twelve agencies including the **FAA** (Federal Aviation Administration), **MARAD** (Maritime Administration), **FRA** (Federal Railroad Administration), **NHTSA** (National Highway Traffic Safety Administration), etc.

$\mathbf{D_R}$ Wheel rim nominal diameter (in).

$\mathbf{\Delta}$ Greek upper case letter "**delta**", usually symbolizing a finite change in the value of something, like a change in lateral force per a change in slip angle: "$\mathbf{\Delta F_y/\Delta\psi}$".

∂ The partial differential symbol indicates an infinitesimal variation in some parameter value.

$\mathbf{\delta}$ Greek lower case letter "**delta**", usually symbolizing the steering angle at a wheel (deg, rad).

$\mathbf{\delta_i}$ The steering angle at the "inner" wheel in a turn (deg, rad).

$\mathbf{\delta_o}$ The steering angle at the "outer" wheel in a turn (deg, rad).

$\mathbf{\delta_{ss}}$ The steady-state "**Steering Angle**", a directional stability metric which is calculated for a turn at some radius "**R**" and lateral acceleration "$\mathbf{a_y}$" (deg, rad).

$\mathbf{d_y}$ A small lateral offset distance, like the lateral shift in tire tread with respect to the tire vertical axis when under a lateral (side) load (in, mm).

e The "pneumatic trail" which is the longitudinal distance in the tire-road contact patch from the patch center point to the line of action of the tire's lateral traction force (in, mm).

EAR The United States "**Export Administration Regulations**".

ESC "**Electronic Stability Control**" is an automotive electronic control system of greater sophistication than **ABS** or even **TCS**. When sensor input indicates that the driver is about to lose control the **ECS** seeks to prevent that loss through modulation of drive and/or braking torques. **ECS** systems developed from the rise of **ABS** but, unlike **ABS** which activates only upon brake application, **TCS** is continually active. There are a number of **ESC** systems existing which have various names such as "**Vehicle Stability Control**" (VSC) or "**Dynamic**

Stability Control" (**DSC**), but they all have the same function although varying in detail.

F A force, as in "**F = ma**" (lb, kg, N). However, in turning "**F**" is the centripetal traction force generated by a tire which may be resolved into components "$\mathbf{F_y}$" (**lateral traction force**) and "$\mathbf{F_x}$" (**cornering drag force**) (lb, N). In the context of this paper it is assumed that the simulation skidpad radius is great enough that such angularity effects may be neglected with "**F**" and "$\mathbf{F_y}$" essentially the same and "$\mathbf{F_x}$" being considered insignificant.

$\mathbf{F_d}$ A directional disturbance force (lb, N).

$\mathbf{F_f}$ The total traction force at a vehicle front axle (lb, N).

$\mathbf{F_y}$ The lateral tire traction component of the "**F**" force in a turning situation (lb, N).

$\mathbf{F_{yaif}}$ The actual lateral tire traction force at the inner front tire in a steady-state turning situation (lb, N).

$\mathbf{F_{yaof}}$ The actual lateral tire traction force at the outer front tire in a steady-state turning situation (lb, N).

$\mathbf{F_{yair}}$ The actual lateral tire traction force at the inner rear tire in a steady-state turning situation (lb, N).

$\mathbf{F_{yaor}}$ The actual lateral tire traction force at the outer rear tire in a steady-state turning situation (lb, N).

$\mathbf{F_{yf}}$ The lateral tire traction force at front tire (axle) in a bicycle model turning situation or Understeer Gradient calculation (lb, N).

$\mathbf{F_{ypif}}$ The potential lateral tire traction force at the inner front tire in a steady-state turning situation (lb, N).

$\mathbf{F_{ypof}}$ The potential lateral tire traction force at the outer front tire in a steady-state turning situation (lb, N).

$\mathbf{F_{ypir}}$ The potential lateral tire traction force at the inner rear tire in a steady-state turning situation (lb, N).

$\mathbf{F_{ypor}}$ The potential lateral tire traction force at the outer rear tire in a steady-state turning situation (lb, N).

$\mathbf{F_{yr}}$ The lateral tire traction force at a rear tire (axle) in a bicycle model turning situation or Understeer Gradient calculation (lb, N).

F_r The total traction force at a vehicle rear axle (lb, N).

FORTRAN "**Formula Translation**" is a general purpose compiled programming language developed by **IBM** in the 1950's that is especially well suited to engineering and scientific applications.

FPS The English (a.k.a. "Imperial") System of Units in feet, pounds, and seconds.

FSAE Stands for the "**Student Formula Competition**" sponsored by the **SAE**.

FWD "**Front wheel drive**", which is a common automotive drive configuration wherein only the two front wheels supply propulsion.

4WD "**Four wheel drive**", which is an automotive drive configuration generally with an option to shift to **RWD**.

F_x The longitudinal tire traction (**drag**) component of the "**F**" force in a turning situation (lb, N).

g The standard value for one gravity unit (32.174 ft/sec^2, 9.807 m/sec^2).

GB Stands for "**Great Britain**", which constitutes England, Scotland, and Wales. When Northern Ireland is included then the nation is referred to as the "**United Kingdom**"

GM See "**GMC**".

GMC **General Motors Corporation**, founded 16 December 1908, was once the largest manufacturing concern in the world, but went bankrupt and was saved by a **US** Government "bail out" in 2009. **GM** is still among the world's largest automotive manufacturers, employing about a quarter of a million people and doing business in 157 nations.

GRC The "**Geometric Roll Center**" or instantaneous center of rotation of a sprung mass as determined by a static kinematic study of the suspension, usually taken as being equivalent to the "**Kinematic Roll Center**" (**KRC**), the "**Force Roll Center**" (**FRC**), and the "**Moment Roll Center**" (**MRC**), although there are distinctions between these roll center types.

GTP "**Grand Touring Prototype**" is a class of **IMSA** race car created in 1981.

h_{rc} The "**height**" of a roll center above (+) or below (-) ground plane (in, cm).

h_{rcf} The "**height**" of the front suspension roll center above (+)/below (-) ground plane (in, cm).

h_{rcr} The "**height**" of the rear suspension roll center above (+)/below (-) ground plane (in, cm).

I/B Inboard, or close to the vehicle longitudinal centerline.

IME The "**Institution of Mechanical Engineers**" (**IME**) is a professional society that was founded by George Stephenson in 1847 at the Queen's Hotel in Birmingham. In 1899 the Institution moved into its permanent headquarters in central London. In 1930 the Institution received a Royal Charter. It has over a 120,000 members in 140 countries, working in many industries including railroad, automotive, aerospace, manufacturing, energy, biomedical, and construction.

IMSA "**International Motor Sports Association**" is a North American sanctioning body for sports car racing that was founded in 1969 by John and Peggy Bishop.

IPS The English (a.k.a. "**Imperial**") System of Units in inches, pounds, and seconds.

IFS Independent front suspension.

IRS Independent rear suspension.

ITAR The United States "**International Traffic in Arms Regulations**".

JCL Mainframe computers such as the **IBM 360** operated in "**batch processing**" environment by way of punched card decks. It was generally required that such decks were to consist of program and data sections sandwiched between "**Job Control Language**" sections. **JCL** identified the resident routines and computer resources to be used for the program run, and was rather "**user unfriendly**".

J-turn A harsh automotive maneuver in which a reversing vehicle is spun 180 degrees and continues in the same direction of travel which it then faces.

K_{roll} The total vehicle (front plus rear suspension) roll stiffness (lb-ft/degree).

K_{froll} The roll stiffness at the front suspension (lb-ft/degree).

K_{rroll} The roll stiffness at the rear suspension (lb-ft/degree).

K_{us} The "**Understeer Gradient**", a quantitative stability metric expressed in degrees or radians per "**g**" ("**$K_{us} > 0$: Stable**", "**$K_{us} = 0$: Neutral**", "**$K_{us} < 0$: Unstable**").

K_z The symbol for a tire's vertical stiffness (lb/in, kg/mm).

L The aerodynamic lift force (lb, kg).

LCG The vehicle **longitudinal center of gravity** location (in, cm).

LCP The vehicle **longitudinal center of pressure** location (in, cm).

L/H Left hand, as opposed to right hand, orientation.

l_f The longitudinal distance from the front axle line to the vehicle **CG** (in, cm). **SAE J670_2008** (para. 4.1.22), uses the symbol "**a**" for this quantity.

l_{fs} The longitudinal distance from the front axle line to the sprung weight **CG** (in, cm).

l_r The longitudinal distance from the rear axle line to the vehicle **CG** (in, cm). **SAE J670_2008** (para. 4.1.22), uses the symbol "**b**" for this quantity.

l_{rs} The longitudinal distance from the rear axle line to the sprung weight **CG** (in, cm).

LS-DYNA A "**Finite Element Modeling**" (**FEM**) based general purpose multi-physics software package developed by the former Livermore Technology Corporation which was acquired by Ansys Inc. in 2019.

LVDS "**Light Vehicle Dynamic Simulation**" is a microcomputer based program that will simulate light ground vehicles undergoing a variety of maneuvers such as cornering, cornering with braking or acceleration, skidding, spinning, and rollover. This program was developed by Andrzej G. Nalecz of the **NHTSA** circa 1992.

l_{wb} The "**wheelbase**", the longitudinal distance between the front and the rear axles (in, ft, cm, m). **SAE J670_2008** (para. 4.1.22), uses the symbol "**L**" for this quantity.

m Mass, as in "**F = ma**" (slugs, kg), <u>or</u> the rate of lateral tire traction potential decrease with load as in the lateral traction coefficient equation: "**μ = b-mN**" (lb^{-1}, when the normal load "**N**" is in lb units).

MATLAB A desktop computer environment ("**mathematics laboratory**") constructed by MathWorks Inc. for iterative analysis and design processes that expresses matrix and array math directly.

m_{if} The coefficient of traction decrease with normal load at the inner front tire (dimensionless).

m_{ir} The coefficient of traction decrease with normal load at the inner rear tire (dimensionless).

MIRA The "**Motor Industry Research Association**" (Ltd) was founded 1949 in the **UK** and is now a major automotive test facility with aerospace, rail, and other branches.

$\mathbf{m_{of}}$ The coefficient of traction decrease with normal load at the outer front tire (dimensionless).

$\mathbf{m_{or}}$ The coefficient of traction decrease with normal load at the outer rear tire (dimensionless).

M/T "Motor Trend" magazine is a major automotive publication (**US**) founded in 1949.

$\mathbf{M_z}$ The tire "**self-aligning torque**" resulting from the "**lateral traction force**" "$\mathbf{F_y}$" and the "**pneumatic trail**" "**e**": $\mathbf{M_z = F_y\, e}$. One might expect the self-aligning torque to increase in step with the lateral traction force and slip angle, but instead the **self-aligning torque** builds to a maximum at around 8 deg **slip angle** and then quickly decays to zero and may go slightly negative (lb-ft, Nm).

N The normal (orthogonal) load bearing on a tire contact patch (lb, N).

NACA The "**National Advisory Committee for Aeronautics**" was founded by the US federal government in 1915 for conducting and promoting aeronautical research. In 1958 it was replaced by **NASA**, the "**National Aeronautics and Space Administration**".

NHTSA The "**National Highway Traffic Safety Agency**" is a major part of the **US DOT**. It describes its mission as "**Save Lives, Prevent Injuries, Reduce Vehicle Related Crashes**". The **NHTSA** is responsible for creating and enforcing "**Federal Motor Vehicle Safety Standards**" (**FMVSS**) and fuel economy regulations.

$\mathbf{N_{of}}$ The normal load on the outer front tire in a steady-state turning situation (lb, N).

$\mathbf{N_{if}}$ The normal load on the inner front tire in a steady-state turning situation (lb, N).

$\mathbf{N_{or}}$ The normal load on the outer rear tire in a steady-state turning situation (lb, N).

$\mathbf{N_{ir}}$ The normal load on the inner rear tire in a steady-state turning situation (lb, N).

NPL In the **UK** the "**National Physical Laboratory**" (**NPL**) was founded in 1900 at Teddington for the purpose of "**standardizing and verifying instruments, for testing materials, and for the determination of physical constants**".

NS The "**neutral steering**" condition of automotive handling.

NSP The "**Neutral Steer Point**" is the automotive analog to the aeronautical "**Neutral Stability Point**".

OAT "**One-At-a-Time**" is one of the simplest and most common methods of Sensitivity Analysis. Each input variable of interest is varied individually while holding all other input variables constant at their nominal values. The variation in output variable per variation in each input variable is noted, and establishes the "**sensitivity**" of the output variable to each input variable. However, if not somehow accounted for, the presence of complex interactions between input variables and highly non-linear relationships can render the sensitivity results erroneous.

o The point of rotation on the tire-road contact area when the tire/wheel is steered.

ω Greek lower case letter "**omega**". Usually stands for an angular velocity (deg/sec). In the steady-state case a turning automobile rotates about a vertical axis through the **CG** with a constant angular velocity.

O/B Outboard, or far from the vehicle longitudinal centerline.

OS The "**oversteering**" condition of automotive handling.

P_i The symbol for tire inflation pressure (psi, kPa).

π The Greek lower case letter "**pi**", generally representing the natural constant "**3.14159……**"

Pro/Engineer A parametric 3D "**Computer Aided Design/Computer Aided Engineering**" (**CAD/CAE**) system running in the Microsoft Windows operating system, and developed by the Parametric Technology Corporation.

ψ Greek lower case letter "**psi**" generally used to symbolize the "**slip angle**" (misnomer) between a tire's direction of movement and the tire's longitudinal axis (degrees, radians)

ψ_f The average "**slip angle**" (misnomer) of the front axle tires (degrees, radians).

ψ_r The average "**slip angle**" (misnomer) of the rear axle tires (degrees, radians).

ψ_{if} The "**slip angle**" (misnomer) of the front inner tire (degrees, radians).

ψ_{of} The "**slip angle**" (misnomer) of the front outer tire (degrees, radians).

ψ_{ir} The "**slip angle**" (misnomer) of the rear inner tire (degrees, radians).

ψ_{or} The "**slip angle**" (misnomer) of the rear outer tire (degrees, radians).

R A radius, such as the radius of a skidpad as measured from the **TC** to the vehicle **CG** (ft, m).

R_i The radius of a skidpad as measured from the **TC** to a vehicle "inner" tire contact patch center (ft, m).

R_o The radius of a skidpad as measured from the **TC** to a vehicle "outer" tire contact patch center (ft, m).

RG The "**roll gain**" is the rate of change of a vehicle's **sprung mass** roll angle "θ_s" per lateral acceleration "a_y" in g's (deg/g, rad/g). It is a linearization of the roll function "$\theta_s = A\, a_y^2 + B\, a_y + C$".

R/H Right hand, as opposed to left hand, orientation.

R&T "**Road & Track**" magazine is an automobile enthusiast's magazine founded at Hempstead, **NY**, **USA** in 1947; it was perhaps the most technically orientated of all major automotive publications.

rpm "**Revolutions per minute**", which might more correctly be termed "**rotations per minute**", generally refers to the angular speed of an engine or motor.

SAE The "**Society of Automotive Engineers**" (now "**SAE International**") was initially founded in 1904 as the "**Society of Automobile Engineers**", mainly as the result of the advocacy efforts of two journalists: Peter Heldt of ***The Horseless Age***, and Horace Swetland of ***The Automobile***. By 1916 the focus was broadened beyond the automobile to include all forms of transportation.

SAWE The "**Society of Allied Weight Engineers**", which is the premier professional society for engineers involved in **mass properties analysis and control**. The "**Society of Aeronautical Weight Engineers**" was organized in 1939 in Los Angeles, California, and was incorporated as a nonprofit organization April 2, 1941. As membership grew to include engineers associated with shipbuilding, land transportation, and other allied industries and technologies, the Society name was changed on January 1, 1973 to the more inclusive "**Society of Allied Weight Engineers, Inc.**".

SCS "**Suspension Control System**" is a general terminology that could mean any one of a number of approaches to attaining an active suspension system, but at present usually means just automatic damping control.

S_H The tire "shoulder" height (in, mm).

Simulink A **MATLAB** based graphical programming environment for modeling, simulating, and analyzing dynamic systems developed by MathWorks.

SM "**Static Margin**" is an indicator for classifying a vehicle's directional stability as stable ("**understeering**"), neutral, or unstable ("**oversteering**"); it is a dimensionless fraction which can be used to locate the **NSP** with respect to the **LCG** ("**NSP = LCG – (SM × l_{wb})**"). Typical values for passenger cars may range between +0.05 to +0.07 (**understeering**).

S_N The tire nominal cross-section width (in, mm).

SolidWorks A solid modeling "**Computer Aided Design**" (**CAD**) and "**Computer Aided Engineering**" (**CAE**) computer program published by ***Dassault Systèmes*** for microcomputer operation within the Microsoft Windows operating system.

SSF The "**Static Stability Factor**" is a rough indication of a vehicle's overturn threshold as it is calculated without reference to roll or other dynamic effects: "**SSF = t / (2 VCG)**" (dimensionless).

STAR-CCM+ A multi-physics (fluid flow, heat transfer, stress, particulate flow, electro-magnetism) "**Computational Fluid Dynamics**" (**CFD**) simulation program acquired and developed further by Siemens Digital Industries Software.

SuspensionSim A simulation program marketed by Mechanical Simulation Corporation (founded 1996) of Ann Arbor MI for quasi-static "**Kinematics and Compliance**" (**K&C**) analysis of suspension systems. System analysis data can then be used in conjunction with **CarSim**, **BikeSim**, or **TruckSim**.

TC The "**Turn Center**" of a cornering maneuver or of a circling about a skidpad. When circling about a skidpad in a steady-state condition the radial line from the **TC** to the vehicle **CG** is orthogonal to the vehicle velocity vector through the **CG**.

TCG The vehicle **transverse center of gravity** (in, cm).

TCS "**Traction Control System**" is generally an extension of existing **ABS** (**Anti-skid Braking Systems**), and is often part of a more all encompassing **ESC** (**Electronic Stability Control**) system. Just as **ABS** is intent upon avoiding wheel lock-up under braking torque, **TCS** is intent upon avoiding wheel spin under accelerating torque. **TCS** can utilize braking torque to modulate the drive torque, but may also have the ability to modulate engine torque. In 1971 Buick introduced an early form of **TCS** called "**Max Trac**" on the Riviera, Estate Wagon, Electra 225, Centurion, and LeSabre models. Cadillac followed in 1979 with their "**Traction Monitoring System**" (**TMS**) on the Eldorado model.

TN A **NACA** designation meaning "**Technical Note**".

t_f The track at the front axle (in, cm).

t_r The track at the rear axle (in, cm).

θ_s Greek lower case letter "**theta**" with "**s**" subscript represents the automotive **sprung mass** roll angle when the vehicle is undergoing a lateral acceleration (deg).

U The "**Under/Over Steer Gradient**" is a directional stability metric obtained by "subtracting the Ackermann steer angle gradient from the reference steer angle gradient on a flat level road in a steady-state condition" (**SAE** J670_JAN2008, para. 11.3.3.1). In other words, this is essentially the same as symbolized by "**K_{us}**" in this paper.

UK The "**United Kingdom**", meaning England, Wales, Scotland, and Northern Ireland, plus three British Crown Dependencies and fourteen British Overseas Territories.

US The "**understeering**" condition of automotive handling, or the "**United States**".

USA The "**United States of America**".

USAF The "**United States Air Force**".

V A velocity (mph, fps, kph, mps). In the context of this paper "**V**" may be the "new" velocity with respect to the "original" velocity "**V_o**" resulting from the application of a disturbance "**F_d**" or a turning "**F**" force.

VCG The vehicle **vertical center of gravity** (in, cm).

V_{char} The "**Characteristic Speed**" is a stability metric for comparing understeering (stable) designs with respect to the exact degree of understeer present (mph, kph).

Vo A tire's original velocity direction prior the application of a lateral disturbance force or a turning force (mph, kph).

V_{crit} The "**Critical Speed**" is a stability metric for determining the speed at which an understeering vehicle can switch from initially controllable (slight oversteer) to very unstable (infinite oversteer) behavior ((mph, kph).

WER "**Weight Estimating Routine**" is a formula for estimating the **weight** of certain specific elements of aircraft design created from recorded historic **weight** data by regression analysis.

W_f The vehicle **weight** load on the front axle (lb, N).

W_r The vehicle **weight** load on the rear axle (lb, N).

W_s The vehicle "**sprung weight**" (lb, N).

W_t The vehicle total **weight** (lb, N).

W_{us} The vehicle "**unsprung weight**" (lb, N).

APPENDIX B – DISTURBANCE AT CG

Automotive directional disturbance forces usually are considered to act directly at a vehicle's **CG**. This is because most vehicle directional disturbances are road inputs. Consider a vehicle traveling on a flat road, then encountering the "disturbance" of a sudden change in the road level (highly exaggerated in Figure B.1). This results in a lateral component of the **weight** vector (disturbance force "$\mathbf{F_d}$") acting through the **CG**:

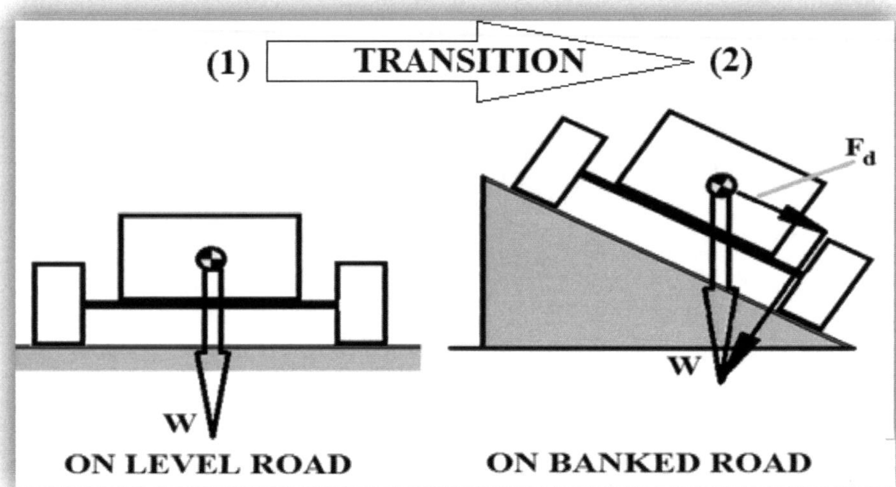

Figure B.1 – ROAD DISTURBANCE

This figure illustrates the major source of lateral disturbance forces, which is road surface undulation (level to banked, bumps, dips, etc.). Wind gusts constitute a secondary but still very significant source of lateral directional disturbance. And, of course, longitudinal and lateral forces inflicted by the road surface at the tires can also impart severe turning moments…

APPENDIX C – DISTURBANCE AT CP

While most disturbance forces act through the **CG**, or generate a reaction force which acts through the **CG**, aerodynamic disturbance forces generally act through the center of pressure or **CP**, which for most modern road vehicles is located just forward of the **CG**. In any case, since all disturbances result in inertial reaction acting laterally though the **CG**, the location of the **CG** is always very significant regarding the nature of the vehicle reaction, but for aero disturbances the **CP** location helps define that reaction.

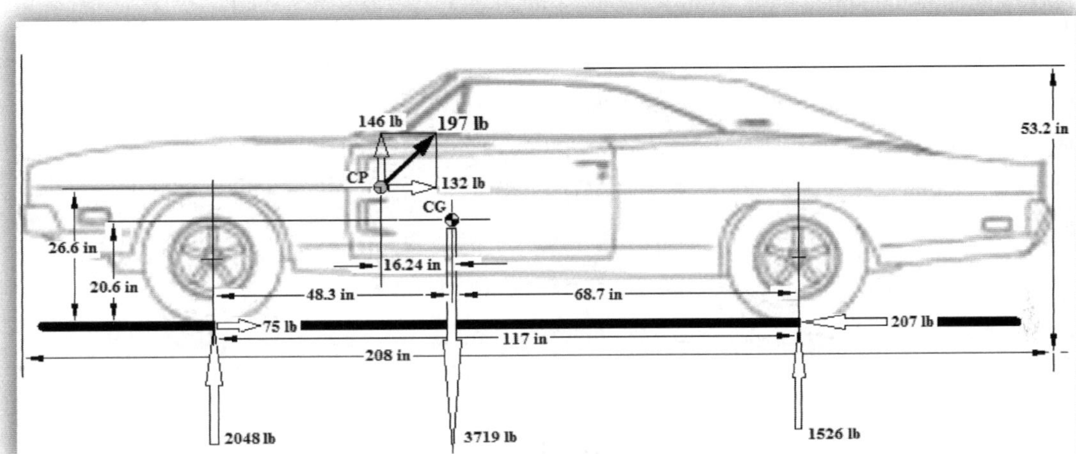

Figure C.1 – AERO LIFT AND DRAG FORCES AT THE CP

The aero forces acting through the **CP** affects automotive directional stability in two ways. The lift and drag forces will modify the normal loads on the tires, and thereby affect the directional stability "indirectly". However, there is also a more "direct" effect, which is commonly expressed:

> ***"If the vehicle becomes airborne, or if the track surface offers little traction, the effect of the contact patches may suddenly disappear or be lessened. In that case, ground vehicle stability, like aircraft stability, requires that the center of pressure be located behind the center of mass. If the vehicle develops a yaw or pitch angle, the aerodynamic forces acting at the center of pressure will be a restoring force if located behind the center of mass, but will tend to increase the yaw or pitch angle if located ahead of the center of mass, possibly leading to a spin or flip. Thus… desirable to have… center of pressure…behind the center of mass."***

Larry Pater[99]

[99] "**Aerodynamics: Drag, Lift, and Stability**", Design and Construction of a Land Speed Record Streamliner, www.paterstreamliner.com, 2016.

Some people may not notice the significance of the words "***airborne***" or "*...**effect of the contact patches may suddenly disappear...***"; there is a lay person assumption that the addition of a large tail fin will augment the stability of any vehicle. Having a **CP** well aft of the **CG** certainly promotes directional stability for an arrow or other body totally dependent on aero forces; but for a wheeled vehicle a lateral aero disturbance force acting through the **CP** location as shown in Fig.C.1, if tire/suspension characteristics are the same all around, that would tend to cause larger slip angles at the front than the rear. That would tend to make the vehicle turn away from the disturbance force such that the resulting centrifugal reaction would be opposing the aero disturbance, tending to even out the front-to-rear slip angle relationship and returning the vehicle to straight ahead motion (!).

For a ground vehicle the stability effect of an aero disturbance depends not just on the exact nature of the **CP/CG** locations, but on those locations with respect to the front and rear tires[100]. Consider the consequences of a lateral aero disturbance for a number of **CG/CP** relationships...

The conventional wisdom is that the **CP** should be aft of the **CG**; usually the **CP** is not far from the **CG** so following two cases (forward **CG** and aft **CG**) quite naturally present themselves as:

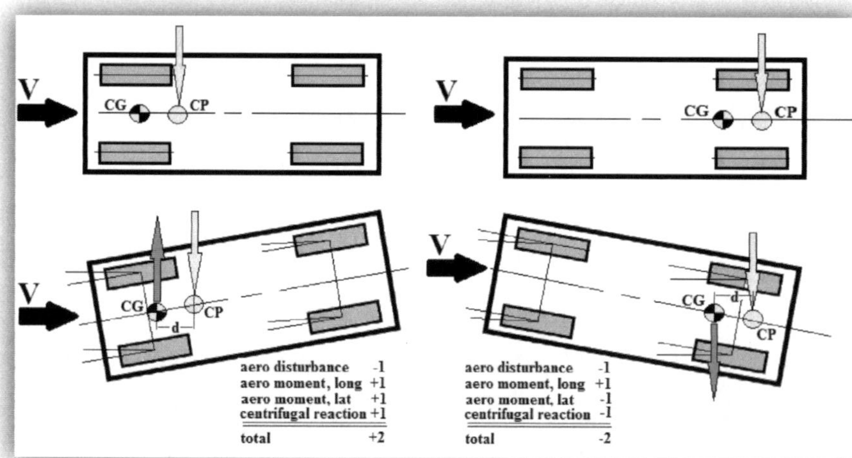

Figure C.2 – CP JUST AFT OF CG, FOR FORWARD CG AND AFT CG

It would seem that, based on this "analysis", the conventional wisdom doesn't work out as well for an aft **CG** car as it does for a forward **CG** car. Now consider the exact opposite of the conventional wisdom, **CP** forward of **CG**, again for the two cases of a forward **CG** and an aft **CG**:

[100] This paper assumes that the lateral aerodynamic disturbance will act at the usual **CP** location as shown in Figure C.1, but a **CP** is much more variable in location than a **CG**. A very strong side gust that is truly orthogonal to the vehicle side profile area will act at the **centroid of that area** which is located just about in line with the driver's seatback in the figure shown.

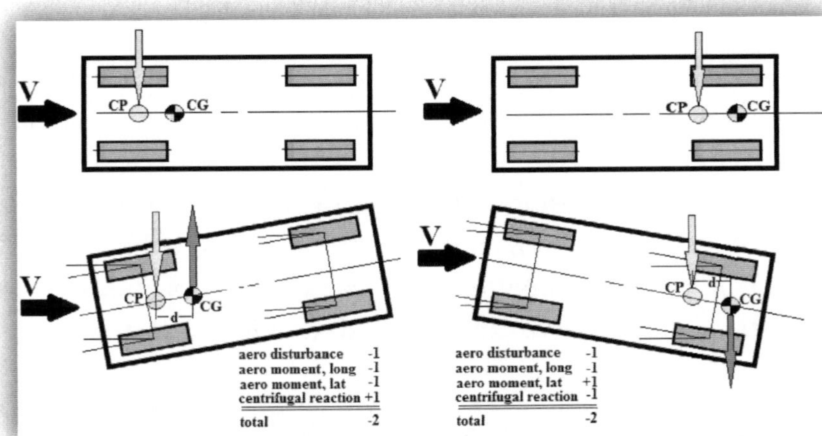

Figure C.3 – CP JUST FORWARD OF CG, FOR FORWARD CG AND AFT CG

Neither of these two cases seems to work out very well with regard to directional stability. However, what happens if the **CG** and **CP** were not so closely coupled? It is unlikely the **CG** and **CP** would ever be quite as divergent as depicted in the next figure, but to complete this "analysis" consider the two cases of a forward **CG**/aft **CP** and a forward **CP**/aft **CG**:

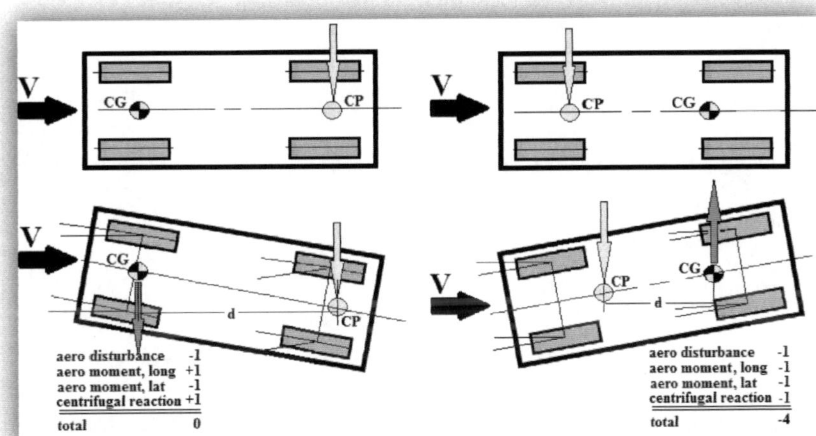

Figure C.4 – FORWARD CG/AFT CP, FORWARD CP/AFT CG

It would seem that a far aft **CP** doesn't work out too badly for a far forward **CG**, but a far forward **CP** with a far aft **CG** works out very, very poorly. Interestingly, early attempts to create ultra-streamlined vehicles of very low drag tended to have very forward **CP's** and very aft **CG's** (due to rear engine placement), leading to comments such as:

"A vehicle in an airstream is the less stable the better its shape is suited for low drag."

Dr. Wunibald Kamm, 1933[101]

[101] Reference [13], pg. 31.

APPENDIX D – OSCILLATORY STABILITY

This paper would not be complete without a word about what Bergman called "**oscillatory stability**". Note a system can be "**statically stable**" but "**dynamically unstable**". A "**statically stable**" system will return to equilibrium after a disturbance input, but in returning may "**overshoot**" the equilibrium position, and then execute a reverse correction back to the equilibrium point, which results in even greater overshoot, etc. Note that although a system may initially exhibit stable behavior, that initial behavior can evolve into an ever increasing series of oscillations; such a system is "**statically stable**", but "**dynamically unstable**".

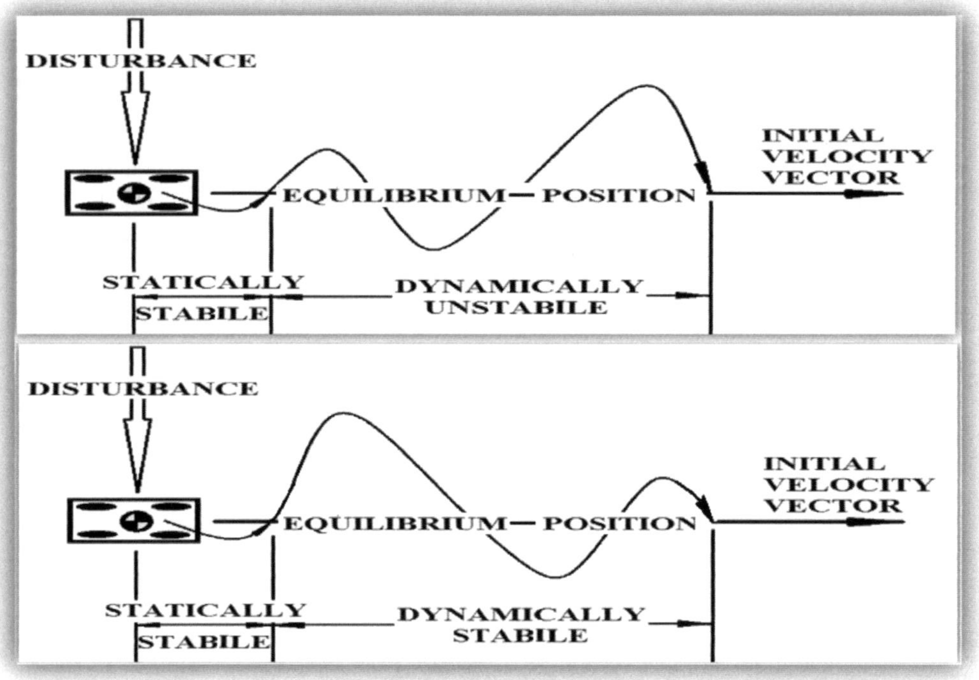

Figure D.1 – STATIC AND DYNAMIC STABILITY

Whether a system is fully, i.e. dynamically, stable or not has to do with the presence of an energy source. The traditional high school physics example of a stable system, such as the one consisting of a marble in a bowl, is limited as that example does not involve a source of energy input; the stability exhibited by the displaced marble in ultimately coming back to rest in the center of the bowl is only a "**static**" stability. Vehicles are quite different; they all have a source of energy input in the form of an engine or motor. The energy input from that engine or motor is in large part reflected in the kinetic energy of the moving vehicle, which is essentially equal to "**$\frac{1}{2}mV^2$**", so it should not be surprising that a full automotive directional stability analysis should hinge on the parameter of "**speed**" (velocity). That is, a vehicle may be dynamically stable in a certain speed range, and unstable in another, usually higher, range. For this paper the involvement of speed in the stability question will only be dealt with as a determinant of the lateral acceleration input to a steady-state analysis.

APPENDIX E – COMPLIANCE STABILITY

Over the course of this paper the concern has been overwhelmingly with the basic, or what Olley would call the "**primary**", stability factors. However, it would be misleading to leave the impression that automotive directional stability is that simple; the matter is complicated by various "**secondary**", or what Bundorf would call "**compliance**", factors. That compliance could influence stability was well understood by Olley's time, but it took Bundorf and Leffert to codify the matter in their paper "**Cornering Compliance Concept for Description of Vehicle Directional Control Properties**" in 1976, which is why a systematic consideration of all the factors influencing a vehicle's directional stability is today called a "**Bundorf Analysis**"[102]:

'05 Porsche Carrera S, 2 Pass., 0.15 g's		FRONT	REAR
	STABILITY FACTORS	(deg/g)	(deg/g)
1	Wt Dist/Load X'fer/Tire	3.20	2.28
2	Camber Thrust	0.09	0.05
3	Roll Steer	0.45	-0.18
4	Align Torque Steer	0.55	0.08
5	Suspension Align Torque Deflect	0.09	-0.04
6	Suspension Lat Force Deflect	0.31	-0.16
	SUMS:	4.69	2.03
	TOTAL K_{us} =	2.66	

Figure E.1 – EXAMPLE BUNDORF ANALYSIS[103]

1) Weight Distribution/Load transfer/Cornering Stiffness of Tire. This most basic and significant contributor to the total **Understeer Gradient** has been pretty well covered in the body of this paper.

2) Camber Thrust. As the vehicle rolls the kinematics of the suspension generally cause a change in the camber of the tires generating forces known as "**camber thrusts**" which have a

[102] Reference [11], pg. 226.

[103] This example is presented courtesy of a Mr. William Cobb, a retired automotive engineer. Note that the example (2005 Porsche Carrera, 2 passengers) has a basic or primary (factor #1) Understeer Gradient of **0.92 deg/g @ 0.15 g's**, whereas the simple spreadsheet upon which this paper is based yielded the basic Understeer Gradient for the 1958 Jaguar XK150S with 2 passengers of **0.62 deg/g @ 0.18 g's**. Study of this example and a few others suggests that factors 1 and 2 are always positive in sign for both front and rear values, but all the other factors can be either positive or negative for rear values.

steering effect. These lateral forces at the tires divided by the associated change in slip angles generate "$\mathbf{C_s}$" values front and rear. Attaining beneficial camber change with suspension motion is a major suspension design concern.

3) Roll Steer. As the vehicle rolls the kinematics of the suspension and steering may cause an unintentional change in the steer angle of the tires generating a steering effect. Such "unintentional" (not the result of driver action) steering effects are usually to be minimized by design, and especially so for race vehicles.

4) Aligning Torque Steer. The tire lateral traction forces produce "**aligning torques**" as shown earlier in Figures 2.1 and 2.2. These torques tend to rotate the vehicle as a whole affecting the slip angle balance front to rear, and therefore the steering. It's easy to see how such torques about the front wheels affect the lateral forces and slip angles at the rear tires, but since the rear tires have slip angles they also have aligning torques, which in turn must affect the lateral forces and slip angles at the front.

5) SAT (Suspension Aligning Torque) Compliance Steer. The aligning torque resulting from the lateral tire traction force (times any pneumatic trail plus caster trail) at the wheel tends to directly twist the suspension due to the built-in compliances, thus affecting the steering and slip angles. Road vehicles tend to have significant aligning torque compliance due to the amount of "**rubber**" used in suspension-to-chassis mounting for vibration and shock attenuation, but race vehicles should have little if any.

6) SLF (Suspension Lateral Force) Compliance Steer. The lateral tire traction force at the wheel tends to deflect the suspension due to its built-in compliances, affecting the steering and slip angles. Road vehicles tend to have significant lateral force compliance due to the amount of "**rubber**" used in suspension-to-chassis mounting for vibration and shock attenuation, but race vehicles should have little if any.

APPENDIX F – TIRE MODEL EQUATIONS

6.00×16 DUNLOP RS4 BIAS TIRE MODEL, SLIP ANGLE & CORNERING STIFFNESS EQUATIONS

FOR N = 0:

$\psi = 0$
$C_s = 0$

FOR N = 200:

$\psi = 0.0000007298\ F_y^3 - 0.0001649884\ F_y^2 + 0.0373856159\ F_y - 0.0569243091$
$Cs = (0.0000021894\ F_y^2 - 0.0003299768\ F_y + 0.0373856159)-1$

FOR N = 400:

$\psi = 0.0000002013\ F_y^3 - 0.0000831348\ F_y^2 + 0.0238845288\ F_y - 0.1077319068$
$C_s = (0.0000006039\ F_y^2 - 0.0001662696\ F_y + 0.0238845288)^{-1}$

FOR N = 600:

$\psi = 0.0000000900\ F_y^3 - 0.0000470133\ F_y^2 + 0.0176483212\ F_y - 0.1059394999$
$Cs = (0.00000027\ F_y^2 - 0.0000940266\ F_y + 0.0176483212)-1$

FOR N = 800:

$\psi = 0.0000000520\ F_y^3 - 0.0000321034\ F_y^2 + 0.0152816867\ F_y - 0.1398755826$
$C_s = (0.000000156\ F_y^2 - 0.0000642068\ F_y + 0.0152816867)^{-1}$

FOR N = 1000:

$\psi = 0.0000000314\ F_y^3 - 0.0000214695\ F_y^2 + 0.0139331283\ F_y - 0.1532355494$
$Cs = (0.0000000942\ Fy2 - 0.000042939\ Fy + 0.0139331283)-1$

FOR N = 1200:

$\psi = 0.0000000181\ F_y^3 - 0.0000128501\ F_y^2 + 0.0130700607\ F_y - 0.1445554791$
$C_s = (0.0000000543\ F_y^2 - 0.0000257002\ F_y + 0.0130700607)^{-1}$

FOR N = 1400:

$\psi = 0.0000000134\ F_y^3 - 0.0000104467\ F_y^2 + 0.0138562141\ F_y - 0.1328149449$
$Cs = (0.0000000402\ Fy2 - 0.0000208934\ Fy + 0.0138562141)-1$

FOR N = 1600:

$\psi = 0.0000000233\ F_y^3 - 0.0000198687\ F_y^2 + 0.0175605669\ F_y - 0.0919030834$
$C_s = (0.0000000699\ F_y^2 - 0.0000397374\ F_y + 0.0175605669)^{-1}$

FOR N = 1700:

$\psi = 0.0000000458\ F_y^3 - 0.0000376466\ F_y^2 + 0.0220521855\ F_y - 0.0917899852$
$Cs = (0.0000001374\ F_y^2 - 0.0000752932\ F_y + 0.0220521855)-1$

FOR N = 1900:

$\psi = 0.000000326337885\ F_y^3 - 0.000191622271760\ F_y^2 + 0.044829711256039\ F_y - 0.030653568375158$
$C_s = (0.000000979013655\ F_y^2 - 0.00038324454352\ F_y + 0.044829711256039)^{-1}$

APPENDIX G - ANGULARITY, RESOLUTION of TIRE TRACTION FORCE(S) into COMPONENTS

The skidpad simulation was written such that the tire traction forces necessary for the attainment of the incremented lateral acceleration levels were calculated without any attempt at the resolution of those forces into longitudinal ("**X**") and lateral ("**Y**") components; the tire traction forces were taken as being essentially the same as the lateral traction forces. The effect of this simplification, in particular upon the effect of the wheelbase ("$\mathbf{l_{wb}}$") parameter in this analysis, can only be understood through a consideration of the geometry of the situation. The following "**bicycle model**" figure is representative of that situation:

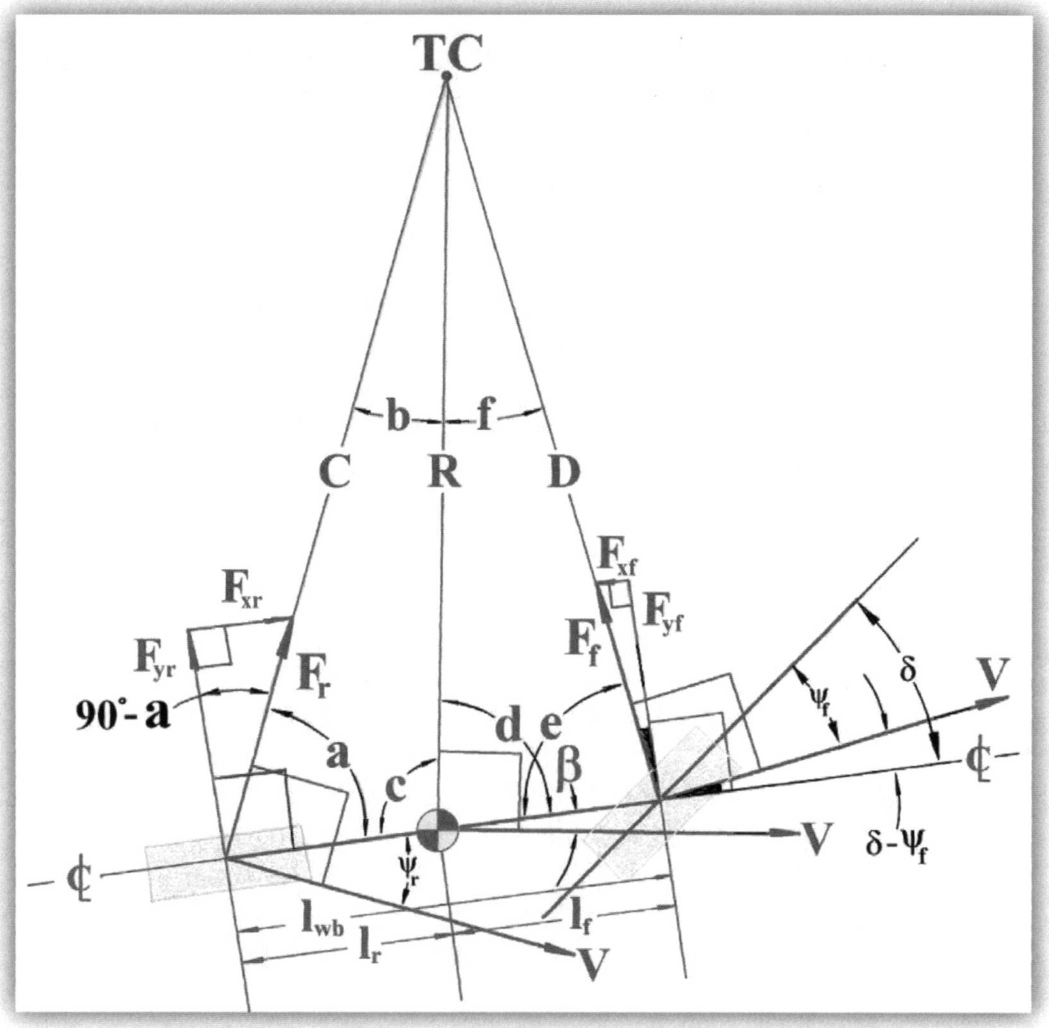

Figure G.1 – GEOMETRY of SKIDPAD SITUATION

The radius of the skidpad ("**R**"), the wheelbase of the subject vehicle ("$\mathbf{l_{wb}}$" or **102.0 in**), and the vehicle **CG** location lengths ("$\mathbf{l_f}$" or **53.01392404 in**, "$\mathbf{l_r}$" or **48.98607596 in**) are the nominal input parameters to the skidpad simulation. The values for "$\mathbf{F_f}$", "$\mathbf{F_r}$", "ψ_f", "ψ_r", and "**δ**" (this last is calculated per the steering angle equation from the "High Speed" Ackermann Steering Geometry, see Fig. 3.5) are output values from the simulation. These values are **671.2 lb**, **726.5 lb**, **2.7426 deg**, **2.88265 deg**, and **4.725 deg** respectively. The remaining values "**a**", "**b**", and "**c**" were calculated per the "**Law of Sines**":

$$\frac{\textbf{Sin }(a)}{A} = \frac{\textbf{Sin }(b)}{B} = \frac{\textbf{Sin }(c)}{C} \qquad \text{(EQ. G.1)}$$

In the context of the situation as depicted in Fig. G.1, the angle "**a**" may be determined:

$$\mathbf{a = 90\ deg - \psi_r = 90\ deg - 2.88265\ deg = \underline{87.11735}\ \underline{deg}}$$

The "**Law of Sines**" may now be more specifically expressed as:

$$\frac{\textbf{Sin }(87.11735)}{1200} = \frac{\textbf{Sin }(b)}{48.98607596} = \frac{\textbf{Sin }(c)}{C}$$

This means that the angle "**b**" can now be determined:

$$b = \sin^{-1}\left(\left(\frac{48.98607596}{1200}\right)\textbf{Sin}(87.11735)\right) = 2.3366\ deg$$

And that the angle "**c**" can also be easily determined given that all three angles ("**a**", "**b**", "**c**") must total 180 deg (they constitute the vertex angles of a triangle):

$$c = 180\ deg - a - b = 180 - 87.11735 - 2.3366 = 90.5461\ deg$$

Lastly, the vehicle sideslip (a.k.a. vehicle attitude) angle "**β**" is calculated per the Ackermann relation (Fig. 3.5)[104]:

$$\beta = 57.3\ {}^{l_r}/_{R} - \psi_r \qquad \text{(EQ. G.2)}$$

Alternatively, the value of "**β**" can be determined from the geometry of Fig, G.1:

$$\beta = 90 - d$$

In the context of the situation as depicted in Fig. G.1, these last two equations may be more specifically expressed as:

[104] The "**57.3**" is a "radians to degrees" units conversion factor.

$$\boldsymbol{\beta} = 57.3\ {}^{48.9861}\!/_{1200} - 2.88265 = \mathbf{-0.543565}\ \boldsymbol{deg}$$

$$\boldsymbol{\beta} = \mathbf{90 - (180 -} \boldsymbol{c}\mathbf{) = 90 - (180 - 90.5461) = 0.5461}\ \boldsymbol{deg}$$

This is as close in absolute value as one could expect for two different "**order of operation**" calculations like this, but it is puzzling about the difference in sign. As drawn in Fig, G.1 the "**β**" angle would definitely be negative, per **SAE** J670_200801[105]. As Eq, G.2 is also an **SAE** standard, it is certain the negative value is correct. For the resolution of tire traction forces "$\mathbf{F_f}$" and "$\mathbf{F_r}$" the significant angles are "**δ-ψ_f**" and "**90-a**", which means the expressions for force resolution are:

$$\boldsymbol{F_{xf}} = \boldsymbol{F_f}\ \mathbf{Sin(\delta - \psi_f)} \quad \text{(EQ. G.3)}$$

$$\boldsymbol{F_{yf}} = \boldsymbol{F_f}\ \boldsymbol{Cos}\mathbf{(\delta - \psi_f)} \quad \text{(EQ. G.4)}$$

$$\boldsymbol{F_{xr}} = \boldsymbol{F_r}\ \mathbf{Sin(90 - a)} \quad \text{(EQ. G.5)}$$

$$\boldsymbol{F_{yr}} = \boldsymbol{F_r}\ \mathbf{Cos(90 - a)} \quad \text{(EQ. G.6)}$$

Substituting the appropriate values for the symbols, the resolution of tire traction forces "$\mathbf{F_f}$" and "$\mathbf{F_r}$" is:

$$\boldsymbol{F_{xf}} = \mathbf{671.2}\ \boldsymbol{lb}\ \mathbf{Sin(4.725 - 2.7426) = 23.2185\ lb}$$

$$\boldsymbol{F_{yf}} = \mathbf{671.2}\ \boldsymbol{lb}\ \boldsymbol{Cos}\mathbf{(4.725 - 2.7426) = 670.7983\ lb}$$

$$\boldsymbol{F_{xr}} = \mathbf{726.5}\ \boldsymbol{lb}\ \mathbf{Sin(90 - 87.11735) = 36.5361\ lb}$$

$$\boldsymbol{F_{yr}} = \mathbf{726.5}\ \boldsymbol{lb}\ \mathbf{Cos(90 - 87.11735) = 725.5807\ lb}$$

So, the simplification of ignoring the angularity in the skidpad simulation means (at 0.384 g's anyway) the difference between using 671.2 lb instead of 670.8 lb, and the difference between using 726.5 lb instead of 725.6 lb. It also means ignoring the net "drag" ("$\mathbf{F_x}$") force of 13.3 lb. Based on this information, it would seem that the decision to ignore angularity in the construction of the skidpad simulation was not a bad move. However, the question remains as to how significant the variation of the wheelbase ("$\mathbf{l_{wb}}$"), say by ±10%, would be to this determination; varying the wheelbase over the range of ±10% produces no significant variation in any of the traction force components, which is the answer to that question.

[105] Reference [35], pg. 17.

APPENDIX H – MAXIMUM LATERAL ACCELERATION vs. PARAMETER VARIATION

The variation in parameter values for this study of directional stability sensitivity to those values also affected the maximum lateral acceleration values:

Effective Ranking	Parameter	Symbol	Units	Maximum Lateral Acceleration (g's) per Variation in Parameter About Norm of 0.659 g: -10%	-5%	0%	+5%	+10%	Comments:
1	Total Vehicle Weight	W_t	lb	0.690	0.674	0.659	0.644	0.629	Reducing total vehicle wt very beneficial for improving max lat accel
2	Sprung Weight	W_s	lb	0.687	0.674	0.659	0.645	0.631	Same, but less wt for same % reduction means slightly less beneficial
14	Unsprung Weight	W_{us}	lb	0.661	0.660	0.659	0.658	0.656	Same, but much less wt for % reduction means much less beneficial
16	Fwd Unsprung Weight	W_{usf}	lb	0.660	0.659	0.659	0.659	0.658	Even less wt for %, and wt effect obscured by consequent LCG effect
17	Rear Unsprung Weight	W_{usr}	lb	0.660	0.660	0.659	0.658	0.658	Even less wt for %, and wt effect obscured by consequent LCG effect
4	Total Vehicle Wt LCG	l_f	in	0.675	0.667	0.659	0.651	0.642	Moving oversteering vehicle LCG fwd tends to improve max lat accel
9	Sprung Weight LCG	l_{fs}	in	0.666	0.663	0.659	0.655	0.651	Same comment as above, but less effect for same % reduction
5	Total Vehicle Wt VCG	h_{cg}	in	0.675	0.667	0.659	0.651	0.642	Reducing VCG is generally always beneficial
6	Sprung Weight VCG	h_{cgs}	in	0.675	0.667	0.659	0.651	0.643	Same comment, but less effective than total vehicle VCG reduction
3	Track	t	in	0.638	0.649	0.659	0.669	0.677	Increasing track (front & rear) generally always beneficial
7	Track Forward	t_f	in	0.646	0.653	0.659	0.665	0.671	Incrs fwd track better than incrs rr track rr for overteering vehicle
10	Track Rear	t_r	in	0.652	0.656	0.659	0.663	0.666	Incrs track rr less good than incrs track fwd as vehicle overteering
18	Roll Axis Height	h_{ra} @ l_{fs}	in	0.660	0.660	0.659	0.659	0.658	Decreasing roll axis ht had a slightly beneficial effect on max lat accel
12	Roll Center Forward	h_{rcf}	in	0.656	0.658	0.659	0.660	0.662	Vehicle had a neg fwd roll center; "increasing" means raising upward
13	Roll Center Rear	h_{rcr}	in	0.662	0.661	0.659	0.658	0.656	Reducing pos rr RC meant beneficial reduction in roll center moment
15	Total Vehicle Roll Stiff	K_{roll}	ft-lb/deg	0.656	0.658	0.659	0.660	0.661	Generally it is always beneficial to increase roll stiffness
8	Roll Stiffness Forward	K_{froll}	ft-lb/deg	0.650	0.654	0.659	0.664	0.668	Increasing fwd roll stiff beneficial to subject oversteering vehicle
11	Roll Stiffness Rear	K_{rroll}	ft-lb/deg	0.666	0.662	0.659	0.656	0.653	Decreasing rr roll stiff beneficial to subject oversteering vehicle
				Maxi Lateral Acceleration Equal or Better Than Nominal					Generally, it seems that what is good for directional stability tends
				Max Lateral Accelration Equal or Less Than Nominal					to also be good for lateral acceleration.

Table H.1 – MAXIMUM LATERAL ACCELERATION vs. 18 PARAMETERS

These maximum lateral acceleration levels represent the data generation "cut-off" points for this paper's skidpad simulation "runs".

[THIS PAGE DELIBERATELY LEFT BLANK]

SAWE Paper No. S3768
Category No. 10
Category No. 11

MASS PROPERTIES REPORTING

Yiyuan Ma, Jin Yan, and Ali Elham
Institute of Aircraft Design and Lightweight Structures
Technische Universität Braunschweig

For Presentation at the
80th SAWE International Conference on Mass Properties Engineering
November 1-4, 2021

Cocoa Beach, Florida

Initial Weight Estimation of Twin-Fuselage Configuration in Aircraft Conceptual Design

Yiyuan Ma[1], Jin Yan[2], and Ali Elham[3]
Technische Universität Braunschweig, 38108 Braunschweig, Germany

Abstract: The Ultra-High Aspect Ratio Wing (UHARW) concept can improve the aircraft's aerodynamic efficiency and reduce fuel consumption. The Twin-Fuselage (TF) configuration is one of the most promising concepts for the UHARW design to reduce the wing bending moments and shear forces. This paper presents the development of a semi-empirical method for the weight estimation of TF aircraft in the initial sizing stage. A physics-based wing weight estimation method is improved for higher aerodynamic analysis fidelity and composite materials, which is used in the design of experiments and the results are applied for regression analysis to establish a semi-empirical method. Eventually, the established semi-empirical weight estimation method is integrated into a TF aircraft conceptual design and performance analysis framework, and a mid-range TF aircraft and a long-range TF aircraft are designed and sized to illustrate its application and efficiency in rapidly estimating the TF aircraft weight breakdown.

Keywords: Twin-fuselage configuration, initial weight estimation, vortex lattice method, composite materials, aircraft conceptual design

1. Introduction

NASA and European Commission have put forward stringent sustainability goals for the next-generation transport aircraft in recent years, including significant reductions in CO2, NOx, noise, etc [1, 2]. In addition, an unexpected revolution in air transportation is needed to recover the aviation industry, which has been affected seriously by the unexpected COVID-19 pandemic, to restore its competitiveness, and to

---

[1] PhD student, Institute of Aircraft Design and Lightweight Structures

[2] Graduate student, Institute of Aircraft Design and Lightweight Structures

[3] Professor, Institute of Aircraft Design and Lightweight Structures

address future climate goals. Accordingly, a lot of research is underway, including novel configurations [3], advanced airframe technologies [4], high-efficient and low-cost design methods [5], etc., to achieve a step-change in aircraft performance. One of the most promising solutions to achieve this purpose is the Ultra-High Aspect Ratio Wing (UHARW) configuration, which can improve aerodynamic efficiency and reduce fuel consumption significantly [6].

However, the bending moment and shear force in the UHARW structure is significantly larger than that of the conventional aircraft wings. The UHARW structure needs to be strengthened to carry the larger loads, which results in a significant increase in the wing weight. Therefore, some novel concepts need to be used to lift the wing bending moment and shear force, so as to maximize the benefits of the UHARW concept.

Robust- and sustainable-by-design ultra-higH aspEct ratio wing and Airframe (RHEA) is a European Union-funded project within the Clean Sky 2 Joint Undertaking (rhea-cleansky2.org/). The RHEA research team, composed of TU Braunschweig (DE), University of Strathclyde (UK), Imperial College (UK), DNW Wind Tunnels (NL), and IRT-Saint Exupery (FR), aims to design next-generation passenger aircraft with ultra-high aspect ratio wings and associated airframe to improve aircraft fuel efficiency. To this end, the overarching objective of RHEA is to improve the aerostructural design and efficiency of UHARW by combining advanced numerical and experimental methods for Multidisciplinary Design and Optimization (MDO).

The twin-fuselage (TF) configuration can significantly reduce the wing bending moment and allow for a lighter wing weight by replacing the large mass of the centrally positioned fuselage with two masses positioned outboard [7]. Besides, since the skin thickness of a pressure cabin is proportional to its volume, a reduction in the fuselage weight can be expected due to the smaller individual fuselage diameter of the TF configuration [8]. Therefore, the TF configuration is considered as one of the solutions to utilize the UHARW design in the RHEA project. The TF configuration has been successfully researched and implemented on large transonic airplanes and unmanned aerial vehicles (UAVs), demonstrating the maturity of related technologies. For example, Scaled Composites developed a TF aircraft Stratolaunch carrier aircraft for an air-launch space access system, which is the largest all-composite aircraft ever built

[9]. Virgin Galactic developed a four-engine TF aircraft WhiteKnightTwn to launch the sub-orbital vehicle SpaceShipTwo [10]. The German Aerospace Center (DLR) has developed and flown a 4-seat TF aircraft HY4, which is powered by a hydrogen fuel cell system [11]. Ma et al. [12] developed and flight-tested a TF aircraft with distributed electric propulsion. Moreover, there are many studies on TF transport aircraft. Lockheed and NASA performed the conceptual design and wind tunnel experiments for a large cargo TF aircraft in the same class as the Boeing 747 transport aircraft and C-5A [13] and conducted the simulator study of its flight characteristics during approach and landing [14]. Chiesa et al. [15] presented a TF aircraft preliminary design study. Vedernikov et al. [16] analyzed the advantages of the TF configuration through a design case based on the prototype of A320.

However, these studies on the TF aircraft design neither introduce the complete design framework and process, nor present the conceptual design methodology, which is challenging to be used as a reference in the TF aircraft design. Besides, weights estimation is particularly important in conceptual design due to its notable impact on the overall aircraft performance. Udin et al. [17] developed a semi-analytical wing mass estimation method for the TF aircraft configuration, which is a class II & 1/2 wing mass estimation method and requires detailed wing design parameters. In contrast, a semi-empirical weight estimation method that requires fewer inputs and has acceptable accuracy is more suitable for the early design stage of TF aircraft. Semi-empirical weight estimation methods are presented in many aircraft design handbooks for conventional aircraft [8, 18]. These methods are either too complex for the initial sizing of the TF aircraft or are applicable only for conventional configurations. Therefore, a semi-empirical weight estimation method for the TF configuration applicable to the initial sizing is required.

To this end, a semi-analytical TF aircraft wing mass estimation method is improved in this work, including improved fidelity of aerodynamic load estimation and extensions to advanced composite materials. Design of Experiment (DoE) and regression methods are used to establish a class II wing mass estimation method for TF aircraft based on the modified physics-based method. Finally, the class II method is integrated into a TF aircraft conceptual design framework and two different classes of TF passenger aircraft are designed to illustrate the design process of the TF aircraft and to demonstrate the application of the developed TF aircraft mass estimation method.

2. Methodology

The weight estimation methodology presented in this section is a combination of the method for the classical aircraft weight estimation method [18, 19] and the semi-empirical wing mass estimation method developed for the TF configuration.

2.1. Initial Weight Estimation

According to the classical aircraft design method [18], the maximum takeoff mass can be expressed as:

$$m_{\mathrm{TO}} = m_{\mathrm{crew}} + m_{\mathrm{pay}} + m_{\mathrm{f}} + m_{\mathrm{e}} \tag{1}$$

where m_{crew} is the crew mass, m_{pay} is the payload mass, m_{f} is the fuel mass, and m_{e} is the empty mass.

In Eq.(1), m_{crew} and m_{pay} can be obtained from the Top-Level Aircraft Requirements (TLAR), and the fuel mass can be calculated by the Breguet range equation which is with respect to the design range from the TLAR. For conventional aircraft, the empty mass is calculated by the empty mass fraction ($m_{\mathrm{e}}/m_{\mathrm{TO}}$), which can be estimated statistically from historical trends developed based on the database. However, this estimation method is only applicable to conventional aircraft and cannot show convincing accuracy for unconventional aircraft, such as the TF configuration, due to the lack of these novel aircraft configuration's real statistical data. Since the most significant difference between the TF aircraft and conventional aircraft is the wing mass estimation method due to the different load distribution, this paper establishes a semi-empirical wing mass estimation method for the TF aircraft applicable to the initial sizing stage, which is presented in subsequent sections.

Since the mass estimation method for other components of TF aircraft is similar to that of conventional aircraft, including fuselage, tailplanes, engines, etc., the classical mass estimation method FLOPS [19] is used. According to the FLOPS, the fuselage mass can be calculated by:

$$m_{\mathrm{f}} = 1.35 \cdot \left(L_{\mathrm{f}} \cdot D_{\mathrm{f}}\right)^{1.28} \cdot \left(1 + 0.05 \cdot N_{\mathrm{e}}\right) \cdot \left(1 + 0.38 \cdot F_{\mathrm{c}}\right) \cdot N_{\mathrm{f}} \tag{2}$$

where L_{f} is the total fuselage length, D_{f} is the average fuselage diameter, N_{e} is the fuselage mounted engines' number, F_c is the cargo aircraft floor factor (0.0 is for passenger aircraft and 1.0 is for military cargo aircraft), and N_f is the fuselages' number.

And the horizontal tail and vertical tail masses can be estimated as:

$$m_{ht} = 0.53 \cdot S_{ht} \cdot DG^{0.2} \cdot (\lambda_{ht} + 0.5) \quad (3)$$

$$m_{vt} = 0.32 \cdot S_{vt} \cdot DG^{0.3} \cdot (\lambda_{vt} + 0.5) \cdot N_{vt}^{0.7} \quad (4)$$

where S_{ht} and S_{vt} are the areas of the horizontal tail and vertical tail, respectively, DG is the designed gross mass, λ_{ht} and λ_{vt} are the taper ratios of the horizontal tail and vertical tail, and N_{vt} is the number of vertical tails.

More details of the other components' mass estimation method, including engines, landing gears, paint, systems, etc., can be found in Ref. [19].

2.2.Wing Mass Estimation Method

This section presents the development of a semi-empirical wing mass estimation method for the TF configuration based on a modified physics-based method.

2.2.1. Improvement of a Semi-Analytical Wing Mass Estimation Method

The most significant difference between TF aircraft and conventional aircraft in the initial sizing is the wing mass estimation approaches due to their different load distribution [16], as illustrated in Fig. 1, which means that the conventional aircraft's wing mass estimation methods do not apply to the TF configuration.

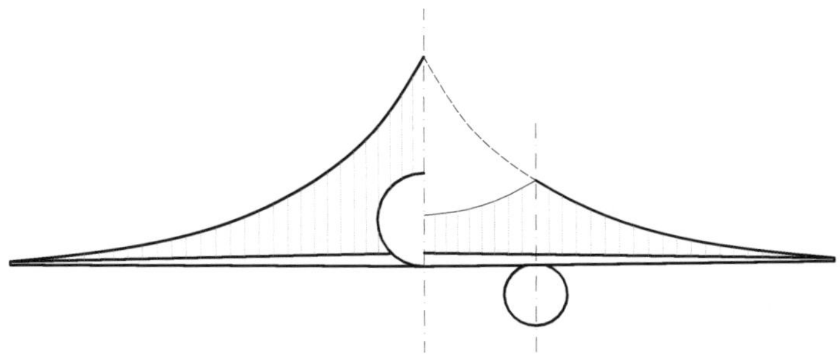

Fig. 1. Wing bending moment comparison between TF configuration (right) and conventional configuration (left) [16].

Udin et al. [17] developed a semi-analytical wing mass estimation method for the TF configuration, estimating the wing structural mass through integrating the wing spanwise mass distribution, including wing structure mass, fuel mass, and concentrated mass such as the engines, with the aerodynamic load. The wing structural relative mass is expressed as

$$m_{\mathrm{s}} = k_{\mathrm{sl}} k_{\mathrm{tw}} k_{\mathrm{man}} \left(m_M + m_Q \right) + m_{\mathrm{rib}} + m_{\mathrm{ail}} + m_{\mathrm{sk}} + m_{\mathrm{flap}} \tag{5}$$

where m_{s} is the relative wing structural mass, k_{sl} is the service life factor, k_{tw} is the twist moment factor, and k_{man} is the manufacturing factor, m_{rib} is the relative rib mass, m_{ail} is the relative ailerons mass, m_{sk} is the relative load free wing skin mass, m_{flap} is the relative flaps mass, and m_M and m_Q are the estimated relative structural mass counteracting the wing bending moment and shear force, which can be calculated as

$$m_M = 2 \frac{\rho n_z g}{\sigma_u T_r} \frac{b^2}{4} E_T \int_0^1 \frac{Mo_{\mathrm{sum}}}{P(y) \cos \Lambda} \mathrm{d}y \tag{6}$$

$$m_Q = \frac{\rho n_z g}{\sigma_{us}} \frac{b}{2} \int_0^1 \frac{Q_{\mathrm{sum}}}{\cos \Lambda} \mathrm{d}y \tag{7}$$

where ρ is the structural material density, n_z is the design load factor, g is the gravitational acceleration, σ_u is the ultimate direct stress, σ_{us} is the ultimate shear stress, b is the wing span, Λ is the wing half-chord sweep, Mo_{sum} and Q_{sum} are the total reduced bending moment and the total reduced shear force caused by aerodynamic loads, wing structural mass, fuel mass, and concentrated mass, which can be obtained by integrating the spanwise distributed loads:

$$Mo(y)=\int_y^1 Q(y)\,\mathrm{d}y \tag{8}$$

$$Q(y)=\int_y^1 q(y)\,\mathrm{d}y \tag{9}$$

where $q(y)$ represents the wing spanwise distributed loads, including aerodynamic load, fuel mass, wing structural mass, and concentrated loads.

For the aerodynamic estimation, a linear equation is used in this method to model the wing spanwise distribution of aerodynamic loads, which is given by:

for the inboard wing section ($0<y<y_{\text{f}}$):

$$q_{\text{a}}=\frac{2}{y_{\text{f}}(1-\lambda)+\lambda+1} \tag{10}$$

and for the outboard wing ($y_{\text{f}}<y<1$):

$$q_{\text{a}}=2\frac{\left[(1-y)/(1-y_{\text{f}})\right](1-\lambda)+\lambda}{y_{\text{f}}(1-\lambda)+\lambda+1} \tag{11}$$

where λ is the wing taper ratio, and y_{f} is the fuselage spanwise location.

The wing secondary structures mass can be estimated by existing semi-empirical methods [8, 17]. Some of them are shown below.

The total aileron mass can be estimated by [8]:

$$m_{\text{ail}}=3m_{\text{ref}}k_{\text{bal}}\left(S_{\text{ail}}/S_{\text{ref}}\right)^{0.044} \tag{12}$$

where m_{ref} is the reference specific weight, which is equal to $56\ \text{N/m}^2$ for Al-alloy skins, S_{ail} is the total platform area of all ailerons, S_{ref} is the wing reference area. The factor k_{bal} is 1.0 for unbalanced ailerons, 1.3 for aerodynamic-balanced ailerons, and 1.54 for mass-balanced ailerons.

The mass of leading-edge flaps can be calculated as:

$$m_{\mathrm{LE}} = 4.83 m_{\mathrm{ref}} \left(S_{\mathrm{LE}} / S_{\mathrm{ref}} \right)^{0.183} \tag{13}$$

where S_{LE} is the leading-edge flaps mass.

The mass of trailing edge flaps can be calculated by:

$$m_{\mathrm{ref}} = 1.7 k_{\mathrm{sup}} k_{\mathrm{slot}} m_{\mathrm{ref}} \left[1 + \left(m_{\mathrm{TO}} / m_{\mathrm{r}} \right)^{0.35} \right] \tag{14}$$

where $m_{\mathrm{r}} = 10^6\ \mathrm{N}$, and the factor k_{sup} and k_{slot} represent the flap motion support and the number of slots, which can be found in Ref. [8].

Besides, miscellaneous items' mass should also be considered, which represents the scattered mass components. Since TF aircraft's landing gears could be stowed underneath the fuselage cabin floor, i.e., no extra fairings are required for landing gears, the mass of the miscellaneous items for TF aircraft should be relatively small, weighing less than 1% of the wing mass [8].

The linear approximation method for lift distribution in the wing box structure is not accurate enough for a physics-based wing mass estimation method. Besides, since this method was published decades ago, it does not apply to advanced composite materials. These shortcomings will be improved in the subsequent sections.

a. Aerodynamic Analysis Fidelity Improvement

Aerodynamic loads account for a remarkable proportion of the total wing loads. Therefore, it is necessary to estimate it accurately, otherwise, the estimated wing mass will have a significant error.

In this work, a VLM tool AVL [20] is integrated into this method to estimate the wing spanwise aerodynamic loads (q_{a}) distribution according to the input wing geometries. By integrating q_{a} along the wing span, one obtains the reduced shear force Q_{a} due to the aerodynamic loads. Then integrating Q_{a} along the wing span, one obtains the reduced bending moment M_{a} due to the aerodynamic loads.

A comparison of the spanwise distribution of q_a, Q_a, and M_a of the VLM and the original linear approximation method is shown in Fig. 2. It can be seen that there is a significant difference in the spanwise aerodynamic loads' distribution between the VLM and the linear approximation method, resulting in different distributions of Q_a and M_a.

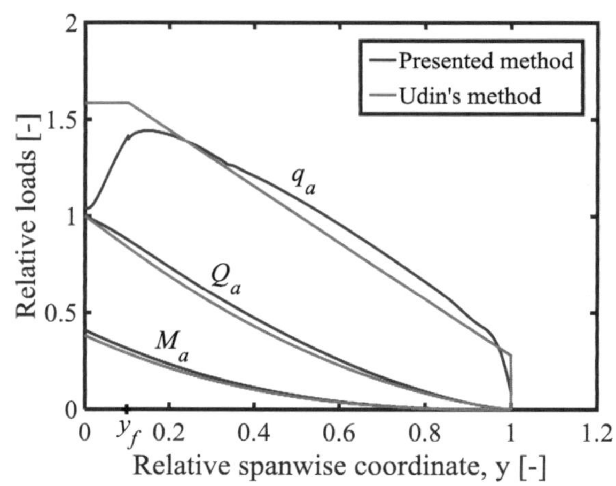

Fig. 2. Spanwise distributions of reduced quantities caused by aerodynamic loads.

b. Improvement for Composite Wings

It is clear that composite materials will be the major material for the next-generation transport aircraft. Therefore, the presented wing mass estimation method needs to be improved. In preliminary design, the cut-off strain method can be used to size the composite wing structures, which are determined according to the worst of all situations [21]. The cut-off strains combined with the 10% rule [22] are used to calculate the allowable for composite materials. Elham et al. [21] presented the laminate properties of composite materials based on the proposed approaches, and the maximum allowable stress in the wing box lower panel can be expressed as

$$\sigma_{\max,l} = E_x \varepsilon_t \tag{15}$$

where E_x is the laminate stiffness in the x-direction and ε_t is the cut-off tensile strain. The maximum allowable stress in the wing box upper panel is given by

$$\sigma_{\max,u} = E_x \varepsilon_c \tag{16}$$

where ε_c is the cut-off compressive strain. Besides, the buckling stress can be calculated by

$$\sigma_{\max,b} = 0.725F\left(E_{x0}^2 Z\right)^{1/4}\sqrt{\frac{P}{L}} \tag{17}$$

And the maximum allowable shear stress can be expressed as

$$\tau_{\max} = G_{xy}\varepsilon_s \tag{18}$$

where G_{xy} is the laminate shear stiffness and ε_s is the cut-off shear strain.

The required laminate properties in the above equations can be calculated by the 10% rule [22], as follows:

$$E_x = \left(0.1 + 0.9\frac{m}{100}\right)E_{x0} \tag{19}$$

$$E_y = \left(0.1 + 0.9\frac{100-m-n}{100}100\right)E_{x0} \tag{20}$$

$$G_{xy} = \left(0.028 + 0.234\frac{n}{100}\right)E_{x0} \tag{21}$$

$$\nu_{xy} = \nu_{yx} = \frac{1}{1+4\left(\frac{100-m-n}{n}\right)} \tag{22}$$

where m is the plies' percentage at 0 degrees, n is the plies' percentage at ± 45 degrees, and E_{x0} is the stiffness of the unidirectional lamina along the principal direction.

c. Iteration Loops

In this semi-analytical wing mass estimation method, a guess value of the wing structural mass is required as one of the initial inputs. For conventional aircraft, there is a well-known 12% rule [23], i.e., the wing mass can be estimated as 12% of the MTOW. The initial input for the wing structural mass can be taken as 12% of the MTOW, and an additional iteration loop is required to compare the calculated wing structural mass with the initial input value. If their difference is less than the convergence criteria, the calculation result of the wing mass is output, otherwise, the wing mass obtained from this iteration is calculated again as the input value until convergence.

d. Validation

The modified semi-analytical wing mass estimation method that improves the accuracy of aerodynamic analysis and extends the composite material model is validated in this section.

The wing mass estimation method presented in this paper is intended for the advanced TF concept. However, so far there are no data available on "real" wing mass for metallic or composite TF aircraft. Nevertheless, the presented physics-based wing mass estimation method is also applicable to conventional aircraft if y_f equals the fuselage radius [17]. In this paper, several metallic medium-range and long-range aircraft and composite medium-range and long-range aircraft were selected for validation, including B737-200 [24], A320-200 [25], B777-200 [26], A330 [27], D8 [1], PFC [28], and B787 [29]. The results are shown in Fig. 3, including results obtained with the original method proposed by Udin et al. [17], the modified method presented in this paper, and the wing mass data from the above-cited references. It can be seen that the accuracy of the modified method fluctuates for the analysis of different aircraft's wing mass, but in general, is more accurate than the original method.

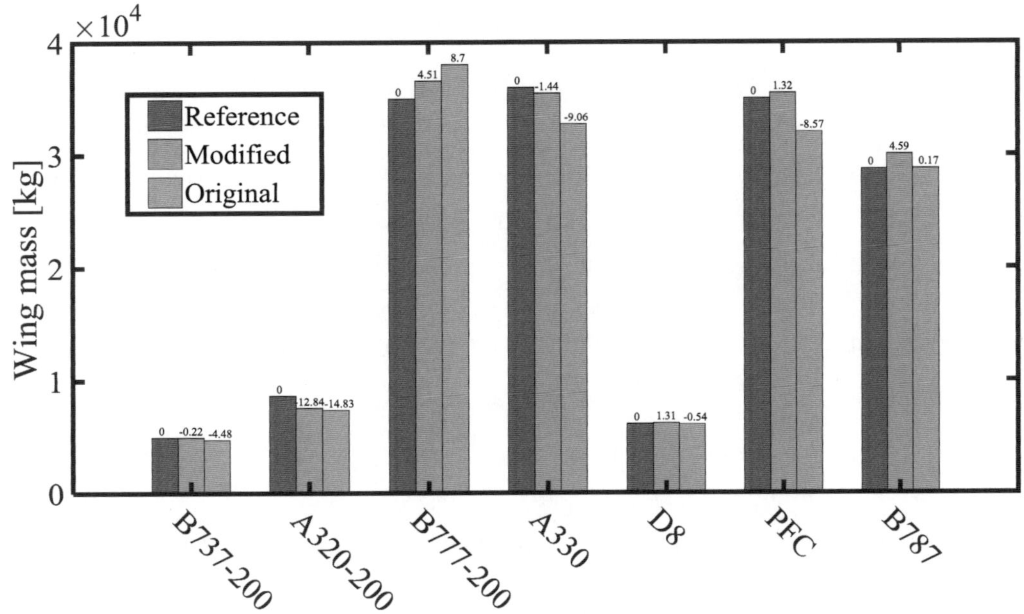

Fig. 3. Validation of medium-range and long-range aircraft with conventional aluminum and CFRP materials ("Original" means the method from Ref. [17]; "Modified" means the presented method in this paper).

2.2.2. Development of a Semi-Empirical Wing Mass Estimation Method

As described, the development of semi-empirical (class II) wing mass estimation methods with sufficient sensitivity to important design parameters is necessary for the conceptual design or initial sizing of unconventional aircraft. In this paper, the semi-empirical methods are developed for four kinds of TF aircraft: MR with metallic materials, LR with metallic materials, MR with composite materials, and LR with composite materials. For each interested configuration, typical design parameters are selected and DoE is performed with the improved physics-based wing mass estimation method. Finally, the estimated wing mass results are used in a multiple linear regression to obtain the semi-empirical equations.

a. Equation Form

The semi-empirical method for the TF aircraft's wing box mass has the following form:

$$m_{\text{wingbox}} = C \cdot m_{\text{TO}}^{E_{\text{m}}} \cdot (W/S)^{E_{\text{ws}}} \cdot AR^{E_{\text{AR}}} \cdot (\cos\Lambda)^{E_{\Lambda}} \cdot (t/c)^{E_{\text{t}}} \cdot V_{\text{m}}^{E_{\text{V}}} \cdot (1+\lambda)^{E_{\lambda}} \cdot n_Z^{E_{\text{nz}}} \cdot (1-Z_{\text{f}})^{E_{\text{zf}}} \cdot Z_{\text{e}}^{E_{\text{ze}}} \quad (23)$$

where the different exponents E and the constant C are used for each configuration, m_{TO} is the maximum take-off weight, W/S is the wing loading, AR is the wing aspect ratio, Λ is the wing quarter-chord sweep angle, t/c is the average wing airfoil relative thickness to chord ratio, V_{m} is the maximum operating velocity (m/s), λ is the wing taper ratio, n_{Z} is the maximum positive load factor, Z_{f} is the relative spanwise fuselage location, and Z_e is the relative spanwise engine location.

In addition, the wing secondary structures mass can be estimated using the semi-empirical methods presented in Sec. 2.2.1.

b. Assumptions

Conventional aluminum structures made from 7075 aluminum are used for the metallic TF aircraft wing box, including covers, spars, webs, and ribs. While the CFRP material is used for the composite TF aircraft wing box, which is built with the graphite-epoxy composite. The wing covers are made of 50% fibers in the center axis direction, 38% fibers in $\pm 45^{\circ}$, and 12% fibers perpendicular to the axis [30]. According to the fiber layup approach, the maximum allowable stresses and shear forces in the wing box panels and webs can be estimated using Eqs. (15-18).

Some typical design parameters are included in Eq. (23), but there are many more parameters influencing the wing mass, including wing platform geometry, structure layout, etc. In order to make the established equations applicable to the early design stage, some assumptions are necessary.

The geometry definition of a TF aircraft wing is shown in Fig. 4. It is assumed that the inboard wing section is not swept and tapered and that the inboard/outboard wing sections are joined at the fuselage centerline. The outboard wing section is tapered, and the engine is assumed to be installed on the outboard wing section.

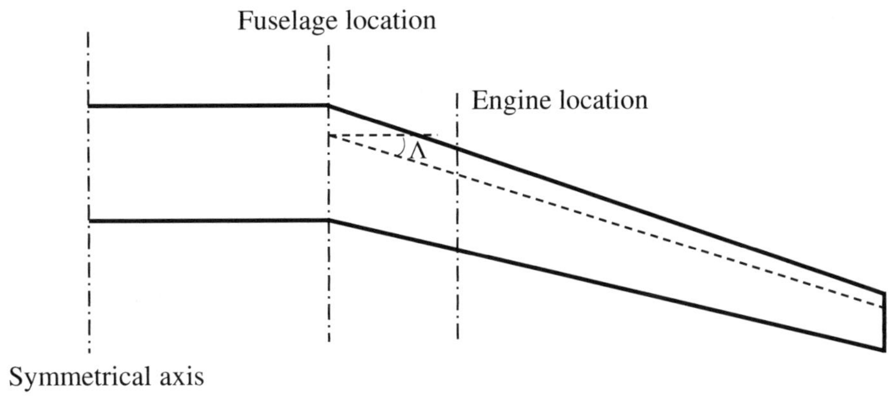

Fig. 4. Wing platform geometry definitions.

The maximum operating Mach number is assumed a function of the outboard wing section sweep angle and wing thickness-to-chord ratio according to the Korn equation [31], as

$$M_{\mathrm{m}} = \frac{k_{\mathrm{a}}}{\cos\Lambda} - \frac{t/c}{\cos^2\Lambda} - \frac{C_{\mathrm{L,des}}}{10\cos^3\Lambda} \tag{24}$$

where k_{a} is the airfoil technology factor, which is taken as 0.93 in this paper and $C_{\mathrm{L,des}}$ is the designed lift coefficient, which is assumed as 0.45 in this paper.

In addition, the ailerons area factor is assumed as 0.06, the flaps area factor is assumed as 0.18, the leading edge flap motion support is assumed to be Fowler flaps with hooked tracks, and the trailing edge flaps are assumed to be double-slotted flaps with articulating vanes.

Most of the presented assumptions are typical for transport passenger aircraft with some minor modifications considering the characteristics of the TF configuration. Some values may be very rough estimates, but this formula is established for the initial sizing purpose providing approximation values for the next stage of high-fidelity wing mass estimation, so these rough estimates are considered acceptable here.

Design of Experiments

DoE is used for regressions to calculate the exponents E and the constant C in Eq. (23). The DoE method of Latin Hypercube Sampling (LHS) is selected to sample the design space since it covers the design space uniformly. One hundred of samples were generated for each configuration, and the lower and upper bounds of the design parameters for the mid-range aircraft and long-range aircraft are listed in Table 1, separately.

Table 1. Design variables boundaries

Design variables	Mid-range aircraft		Long-range aircraft	
m_{TO}, kg	40,000	80,000	200,000	300,000
W/S, (N/m^2)	4,500	6,500	6,500	8,000
AR	15.0	30.0	15.0	30.0
Λ, deg	10.0	25.0	15.0	30.0
t/c	0.1	0.16	0.1	0.16
V_m, (m/s)	200	260	200	260
λ	0.2	0.4	0.2	0.4
n_Z	1.5	2.5	1.5	2.5
Z_f	0.1	0.2	0.1	0.2
Z_e	0.3	0.4	0.3	0.4

c. Regression Analysis

The 100 DoE samples are calculated for each configuration and material, including the metallic MR-TF aircraft, the metallic LR-TF aircraft, the CFRP MR-TF aircraft, and the CFRP LR-TF aircraft, with the improved physics-based wing mass estimation method. The multiple linear regression was used to obtain the exponents and constants in Eq. (23), and the results are given in Table 2.

In the application of this method, the corresponding data from Table 2 can be selected and substituted into Eq. (23) for the analysis according to the mission type and the structural material.

Table 2. Wing box constants and exponents for each concept and material

Aircraft		C	E_{m}	E_{ws}	E_{AR}	E_{Λ}	E_{t}	E_{V}	E_{λ}	E_{nz}	E_{zf}	E_{ze}
Al	MR	-12.8809	1.4487	-0.3817	1.3147	-1.0650	-0.8390	0.1452	2.4358	0.9128	2.6202	-0.1853
	LR	-11.5310	1.3841	-0.3424	1.1530	-0.6774	-0.7501	0.1554	2.3890	0.8221	2.5165	-0.1891
CFRP	MR	-13.1132	1.4732	-0.4303	1.3034	-1.0493	-0.8155	0.1359	2.2995	0.9464	2.5861	-0.1729
	LR	-11.9976	1.4127	-0.3628	1.1639	-0.6991	-0.7396	0.1482	2.3003	0.8671	2.5298	-0.1800

d. Validation

The semi-empirical wing mass estimation method developed here is intended for advanced TF aircraft. There is no real data on the TF aircraft. However, the presented wing mass estimation method is also applicable to conventional aircraft and it is possible to validate the wing mass estimation method with data from conventional aircraft. The selected validation data consists of 11 transport aircraft with aluminum or composite constructions.

The validation results are shown in Fig. 5. Elham et al. [21], Lissys Ltd [32], and Peter et al. [28] presented the wing mass data for the A300, B787, and PFC, respectively. While wing mass data for other commercial transport aircraft were extracted from Ref. [33]. It can be seen that the estimation differences (absolute errors) are below 10% for most aircraft and below 15% for all aircraft, which indicates that the accuracy of the presented semi-empirical method developed for the initial sizing of TF aircraft is acceptable.

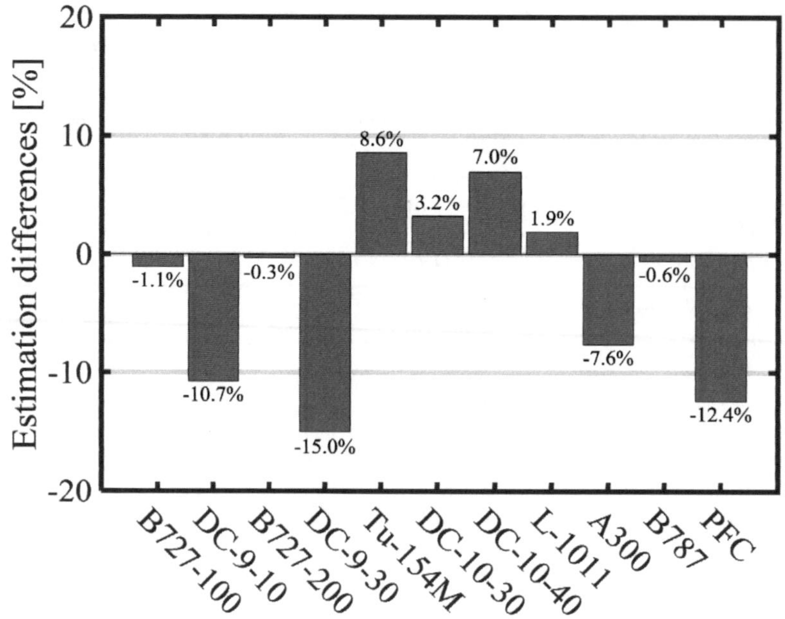

Fig. 5. Validation of the semi-empirical wing mass estimation method for TF aircraft.

e. Impact of Design Parameters

Since the design parameters included in Eq. (23) were selected based on experience and references, sensitivity analysis is used to analyze the importance of each design parameter on the wing mass in order to understand the characteristics of the TF aircraft wing.

The sensitivity analysis in this paper is carried out using a standardized dimensionless form of sensitivity [12]:

$$S_i = \frac{\left(\partial y(X) / y(X)\right)}{\left(\partial x_i / \partial x_i\right)} \quad \text{i= 1, 2, ..., } l \tag{25}$$

A320 and B777 were used as design cases for sensitivity analysis for mid-range and long-range missions, separately. The results are shown in Figs. 6 and 7. Since the same method is used, the importance distribution of each design parameter for MR and LR is basically the same. Maximum takeoff mass and aspect ratio are the most important design parameters because they are the overall parameters that have a significant impact on the wing area. Next are wing thickness-to-chord ratio and maximum positive load factor, which have a strong influence on the wing structural mass, while wing loading, taper ratio, and spanwise fuselage location have similar importance for wing box mass.

The sensitivity analysis used here helps researchers identify the most important parameters by comparing the effects of each design parameter. For example, when sizing a TF aircraft wing, one should realize that the maximum takeoff mass and aspect ratio are the most important parameters and that more attention should be paid to them in order to reduce wing weight.

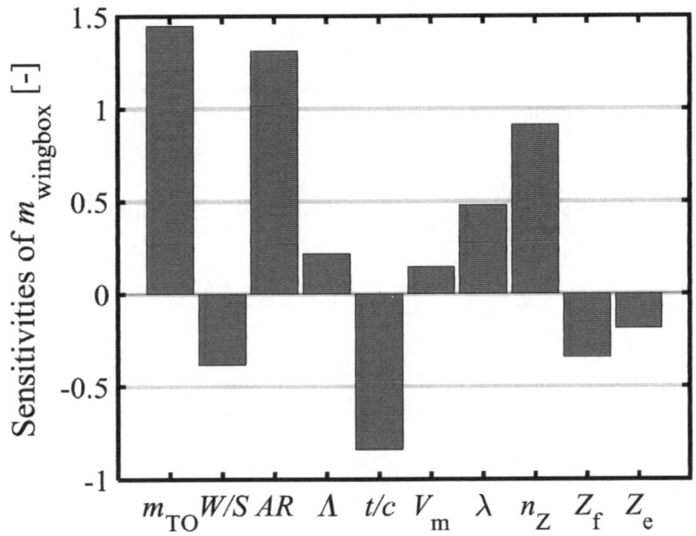

Fig. 6. Sensitivity analysis of wing box mass for mid-range aircraft (A320).

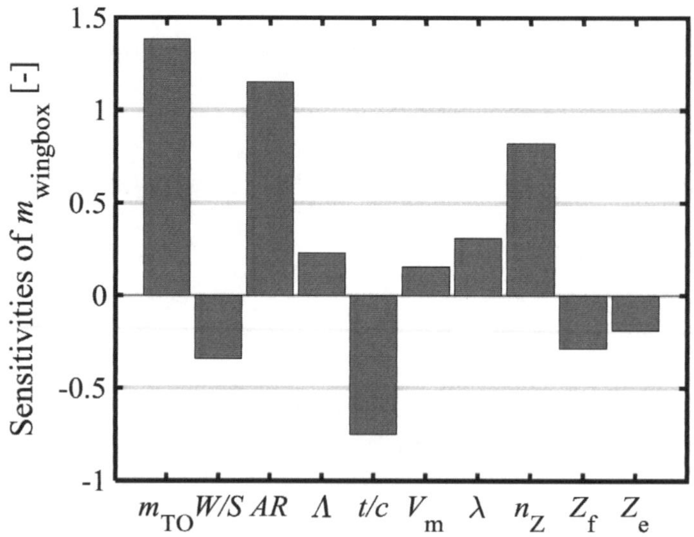

Fig. 7. Sensitivity analysis of wing box mass for long-range aircraft (B777).

2.3. Conceptual Design Framework

A conceptual design and analysis process for TF aircraft has been developed by the authors in Ref. [34], consisting of two main modules: initial sizing and performance analysis. The aircraft initial sizing is performed by using PyInit [6], an in-house tool developed by the authors, to generate the constraints diagrams, size the components, etc. Then the initial sized aircraft is imported into the SUAVE [5], an open-source aircraft performance analysis tool developed by Stanford University, for the analysis of weight breakdown, mission segments, flight performance, etc., through convergence iterations. The semi-empirical wing mass estimation method developed in this paper is integrated into the TF aircraft conceptual design framework, as shown in Fig. 8. The wing mass estimation method in SUAVE is replaced by the presented semi-empirical method for iterative calculations.

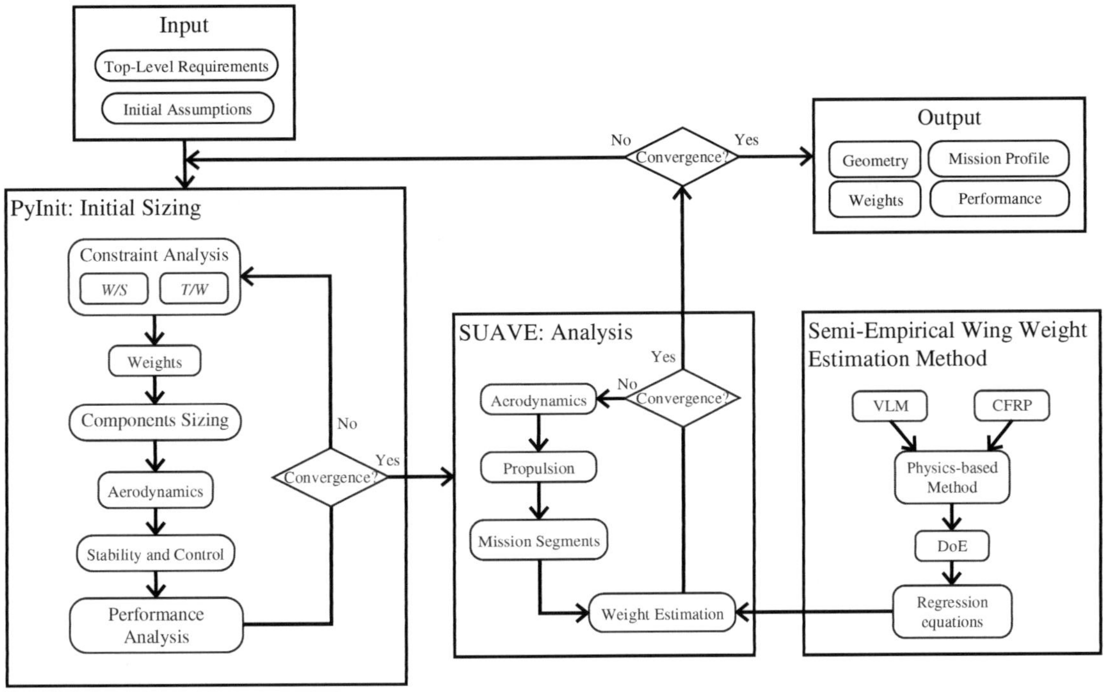

Fig. 8. TF aircraft conceptual design flowchart integrating the developed wing mass estimation method.

3. Design Study

Two design cases, including a Mid-Range (MR) TF aircraft and a Long-Range (LR) TF aircraft, are designed in this section to demonstrate the developed wing mass estimation method and the modified TF aircraft conceptual design framework.

3.1. Top-Level Aircraft Requirements and Assumptions

An MR and an LR twin-fuselage transport aircraft comparable to the A320 and B777 are considered here, separately. The entry-into-service time of these two aircraft is assumed to be 2040, and therefore some novel technology assumptions are introduced, including an assumed 55% laminar flow region of the wing and tails due to hybrid laminar flow control technologies, an assumed 20% reduction in structural mass due to advanced composite materials and structures, and an assumed +1.5g maximum positive load factor due to load alleviation technologies. It should be noted that according to the ICAO airport operation requirements, the wing span and the main landing gear span cannot exceed 36 m and 9 m for the MR-TF aircraft, and 65 m and 14 m for the LR-TF aircraft. The top-level aircraft requirements are given in Table 3.

Table 3. Top-level aircraft requirements

Parameter	Unit	MR-TF	LR-TF
Reference	–	A320	B777
Cruise Mach number	–	0.78	0.84
Cruise altitude	ft	33,000	35,000
Range	nm	3400	7500
Passengers (2 class)	–	150	350
Approach speed	kt	136	140
Diversion range	nm	200	200
Diversion hold	min	10	10
Contingency fuel	%	3	3
Gate-box limit	m	36	65
Main landing gear span	m	9	14

3.2.Sizing and Performance Analysis

Corresponding to the presented technology assumptions and the top-level aircraft requirements, the constraints diagrams of the MR-TF aircraft and the LR-TF aircraft were sized by PyInit. The constraints diagrams and the selected design points are shown in Fig. 9 and 10.

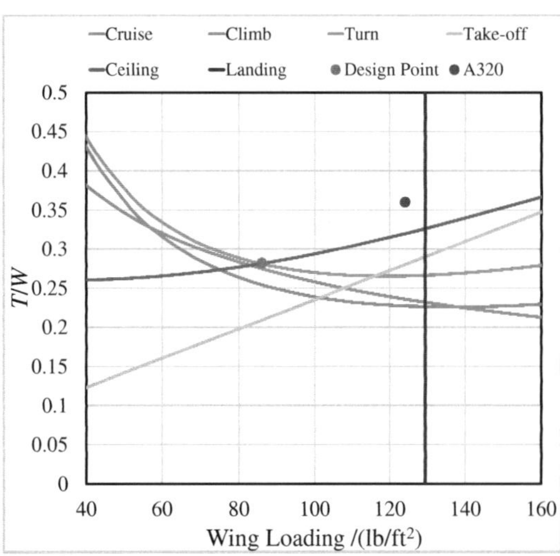

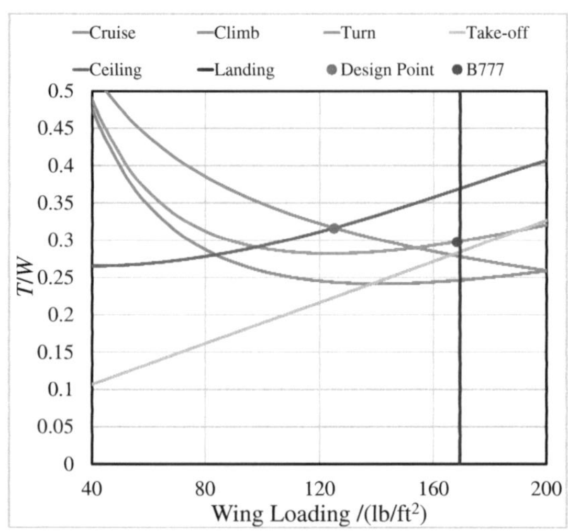

Fig. 9. Constraints diagrams of the MR-

Fig. 10. Constraints diagrams of the LR-

TF aircraft. TF aircraft.

Since the TF configuration has the advantage of significantly reducing the bending moment in the wing structure, these two TF aircraft were designed with ultra-high aspect ratio wings to improve aerodynamic efficiency and reduce fuel consumption. Therefore, the high wing configuration was selected to meet the required engine and wingtip clearance. Considering the aeroelastic characteristics and avoiding the downwash of the wing and engine outflow, the high slab tail configuration was adopted. The geometric dimensions of the initial sized aircraft are shown in Fig. 11 and 12.

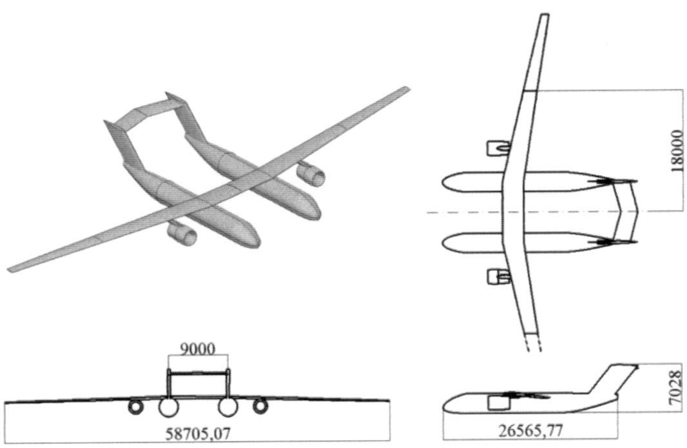

Fig. 11. Geometric dimensions of the MR-TF aircraft.

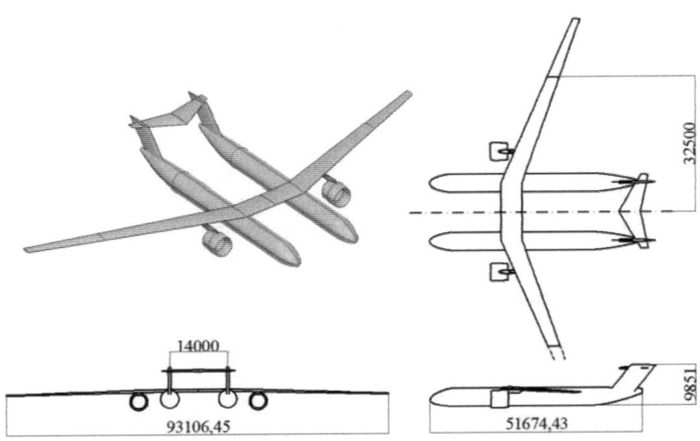

Fig. 12. Geometric dimensions of the LR-TF aircraft.

As illustrated in Fig. 8, the initial sized aircraft were input into the modified SUAVE for performance analysis and assessment with the presented wing mass estimation method. After iterative calculations until convergence, the weight breakdown and mission segments were obtained, as shown in Table 4 and Fig. 13.

Table 4. Weight breakdown summary of the MR-TF and LR-TF aircraft

Parameter	MR-TF	LR-TF
Max. takeoff weight, kg	59,218	243,926
Max. zero fuel weight, kg	45,737	157,328
Fuel weight, kg	13,480	86,597
Empty weight, kg	31,517	124,148
Empty weight breakdown		
Wing, kg	5,755	39,323
Fuselages, kg	5,241	20,596
Propulsion, kg	3,798	17,292
Nacelles, kg	495	2,392
Landing gear, kg	2,021	6,554
Horizontal tail, kg	775	1,565
Vertical tail, kg	850	2,559
Paint, kg	418	1,211
Systems, kg	12,164	32,657

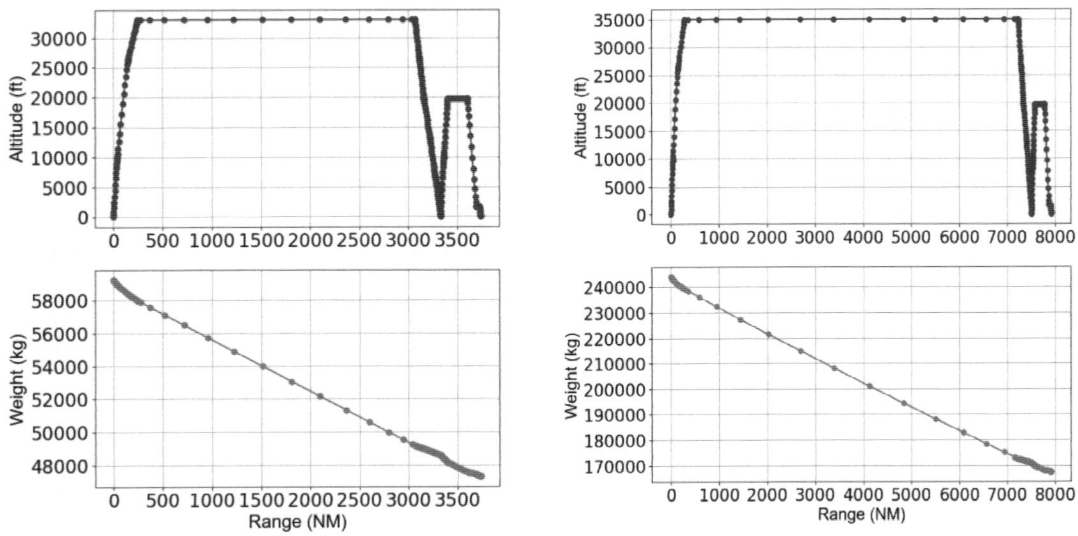

a) MR-TF b) LR-TF

Fig. 13. Mission profiles of the MR-TF and LR-TF aircraft.

4. Conclusion

This paper addressed the initial weight estimation of twin-fuselage aircraft at the conceptual design stage. A semi-analytical wing mass estimation method for the TF configuration was improved in this work, including improved fidelity of the aerodynamic analysis and extensions for advanced composite materials. Since the semi-analytical method requires a large number of input parameters, which are difficult to obtain at the initial sizing stage, DoE and regression methods were used to establish a semi-empirical wing mass estimation method for the TF configuration with different parameters' values for different missions and materials, separately. Eventually, the established semi-empirical weight estimation method was integrated into a TF aircraft conceptual design and analysis procedure and two TF aircraft, including a mid-range TF aircraft and a long-range TF aircraft, were designed and sized to demonstrate the TF aircraft design process and its weight estimation method.

The case study showed that the established initial weight estimation method for the TF aircraft configuration can efficiently analyze the weight breakdown of the TF aircraft within the presented TF aircraft sizing and analysis framework. Since the TF configuration is suitable for utilizing UHARW design, avenues for future work include improving the semi-analytical wing mass method to account for the aeroelasticity and flutter characteristics to develop a semi-empirical wing mass estimation method that is also sensitive to the UHARW characteristics.

Acknowledgments

This project has received funding from the Clean Sky 2 Joint Undertaking (JU) under grant agreement No 883670. The JU receives support from the European Union's Horizon 2020 research and innovation programme and the Clean Sky 2 JU members other than the Union.

References

[1]. Greitzer, E. M., P.A. Bonnefoy, E. de la Rosa Blanco, C.S. Dorbian, M. Drela, D.K. Hall, , R.J. Hansman, J.I. Hileman, R.H. Liebeck, J. Lovergren, P. Mody, J.A. Pertuze, S. Sato, Z.S. Spakovsky, C.S. Tan, J.S. Hollman, J.E. Duda, N. Fitzgerald, J. Houghton, J.L. Kerrebrock, G.F. Kiwada, D. Kordonowy, J.C. Parrish, J. Tylko, E.A. Wen, and W.K. Lord, N+3 aircraft concept designs and trade studies final report, NASA CR-2010-216794/VOL2, 2010.

[2]. Flightpath 2050-Europe's vision for aviation: advisory council for aeronautics research in Europe, EUROPEAN COMMISSION: Brussels, Belgium, 2011.

[3]. Karpuk, S., and A. Elham, "Conceptual design trade study for an energy-efficient mid-range aircraft with novel technologies", AIAA Scitech 2021 Forum, January 2021.

[4]. Sanjay Lulekar, S., P. Ghassemi, H. Alsalih, and S. Chowdhury, "Adaptive-Fidelity Design Automation Framework to Explore Bioinspired Surface Riblets for Drag Reduction", AIAA Journal, Vol. 59 No. 3, (2021) pp. 880-892.

[5]. Lukaczyk, T. W., A.D. Wendorff, M. Colonno, T.D. Economon, J.J. Alonso, T.H. Orra, and C. Ilario, "SUAVE: an open-source environment for multi-fidelity conceptual vehicle design", 33rd AIAA Applied Aerodynamics Conference, June 2015.

[6]. Ma, Y., S. Karpuk, and A. Elham, "Conceptual design and comparative study of strut-braced wing and twin-fuselage aircraft configurations with ultra-high aspect ratio wings", AIAA AVIATION 2021 FORUM, 2021.

[7]. Jong, T., and R. Slingerland, "Analysis of the twin-fuselage configuration and its H-cabin derivative", AIAA's 3rd Annual Aviation Technology, Integration, and Operations (ATIO) Forum, November 2003.

[8]. Torenbeek, E., Advanced aircraft design: conceptual design, analysis and optimization of subsonic civil airplanes, John Wiley and Sons Ltd, West Sussex, 2013.

[9]. Corda, S., C. Longo, and Z. Krevor, "Stratolaunch air-launched hypersonic testbed", 22nd AIAA International Space Planes and Hypersonics Systems and Technologies Conference, September 2018.

[10]. Guerster, M., and E.F. Crawley, "Dominant suborbital space tourism architectures", Journal of Spacecraft and Rockets. Vol. 56 No. 5 (2019) pp. 1580-1592.

[11]. Kallo, J., "DLR Leads HY4 project for four-seater fuel cell aircraft", Fuel Cells Bulletin. 2015 Vol.

11 No. 13 (2015).
[12]. Ma, Y., W. Zhang, T. Zhang, X Zhang, and Y. Zhong, "Sizing method and sensitivity analysis for distributed electric propulsion aircraft", Journal of Aircraft. Vol. 57 No. 4 (2020) pp. 730-741,
[13]. Moore, J. W., E.P. Craven, B.T. Farmer, J.F. Honrath, R.E. Stephens, C.E. Bronson Jr, R.T. Meyer, and J.H. Hogue, Multibody aircraft study, volume I, NASA-CR-165829-VOL-1, 1982.
[14]. Grantham, W. D., P.L. Deal, G.L. Keyser Jr, and P.M. Smith, Simulator study of flight characteristics of a large twin-fuselage cargo transport airplane during approach and landing, NASA Technical Paper 2183, 1983.
[15]. Chiesa, S., M. Di Sciuva, and P. Maggiore, "The double-fuselage layout: a preliminary case study of a possible way of reducing the development costs for new high-capacity aircraft", Proceedings of the Institution of Mechanical Engineers, Part G: Journal of Aerospace Engineering. Vol. 214 No. 2 (2000) pp. 85-95.
[16]. Vedernikov, Y. V., V.E. Chepiga, V.P. Maslakov, E.A. Kuklev, and V.G. Gusev, , "Configuration of the medium-haul twin-fuselage passenger aircraft", International Journal of Applied Engineering Research. Vol. 12 No. 4 (2017) pp. 414-421.
[17]. Udin, S. V., and W.J. Anderson, "Wing mass formula for twin fuselage aircraft", Journal of Aircraft. Vol. 29 No. 5 (1992) pp. 907-914.
[18]. Raymer, D., Aircraft design: a conceptual approach, 6th Edition, AIAA Education Series, American Institute of Aeronautics and Astronautics, Washington, D.C, 2018.
[19]. Wells, D. P., B.L. Horvath, and L.A. McCullers, The flight optimization system weights estimation method, NASA/TM-2017-219627/Volume I, 2017.
[20]. Drela, M., and H. Youungren, "Athena Vortex-Lattice Method", 2020. URL http://web.mit.edu/drela/Public/web/avl/.
[21]. Elham, A. and M.J.L. van tooren, "Effect of wing-box structure on the optimum wing outer shape", Aeronautical journal, 2014. 118 (1199): pp. 1-30.
[22]. Douglas Aircraft Company, M.D.C., "The ten-percent rule for preliminary sizing of fibrous composite structures", SAWE 2054. 1992.
[23]. Elham, A., "Weight Indexing for Multidisciplinary Design Optimization of Lifting Surfaces," PhD Thesis, Delft University of Technology, 2013.
[24]. Boeing Commercial Airplanes, 737 Airplane Characteristics for Airport Planning. Sept. 2013.
[25]. Airbus S.A.S., A320 Aircraft Characteristics Airport and Maintenance Planning, Issue: 30 Sept. 1985, Rev: 1 April 2020.
[26]. Boeing Commercial Airplanes, 777-200LR / -300ER / -Freighter Airplane Characteristics for Airport Planning, D6-58329-2. Rev: May 2015.
[27]. Airbus S.A.S., A330 Aircraft Characteristics Airport and Maintenance Planning, Issue: 1 Jan. 1993, Rev: 1 July 2021.
[28]. Peter, F., J. Bijewitz, and A.L. Habermann, "Definition of the CENTRELINE reference aircraft and power plant systems", 9th International EASN Conference, Athens, Greece (2019) pp. 3-6.
[29]. Boeing Commercial Airplanes, 787 Airplane Characteristics for Airport Planning. 2014.
[30]. P. Chiozzotto, G., "Initial weight estimate of advanced transport aircraft concepts considering aeroelastic effects", AIAA Paper 2017-0009, Jan. 2017.

[31]. Mason, W., "Analytic models for technology integration in aircraft design", Aircraft Design, Systems and Operations Conference (1990) pp. 3262.
[32]. Lissys Ltd, Boeing 787 -8 (Dreamliner) sample analysis. 2005. URL: https://www.lissys.uk/samp1/index.html.
[33]. Beltramo, M. N., D. L. Trapp, B.W. Kimoto, and D.P. Marsh, "Parametric Study of Transport Aircraft Systems Cost and Weight," NASA CR151970, 1977.
[34]. Ma, Y., A. Elham, "Twin-fuselage configuration for improving fuel efficiency of passenger aircraft", Aerospace Science and Technology, (2021) 107000.

Yiyuan Ma received the B.S. and M.S. degrees from the School of Aeronautics, Northwestern Polytechnical University, Xi'an, China, in 2017 and 2020, respectively. He is currently pursuing the Ph.D. degree with the Institute of Aircraft Design and Lightweight Structures, Technische Universität Braunschweig, Germany. His research interests include unconventional aircraft conceptual design, distributed electric propulsion, and multidisciplinary design and optimization.

SAWE Paper No. 3771
Category No. 3, 13

A Look at Inclining Experiment Heel Angles: Measurement Tools and Sensitivity

David Tellet

For Presentation at the
2021 Tech Fair
of the Society of Allied
Weight Engineers, Inc.

November, 2021

Abstract

An inclining experiment is used to indirectly measure the vertical center of gravity of a ship by measuring resultant heel angles for a given weight moved athwartships. The methods for measuring these angles are considered tried and true even though the uncertainties of their accuracy and precision are not well understood. This paper explores the impact of errors from traditional inclining methods and compares them with modern methods. The paper looks at a simulated inclining experiment and explores the change in results when errors are introduced into the measurements. It then looks at electronic inclinometers used in an actual inclining and discusses how that data can be analyzed and how that might affect the results of the experiment. Finally the paper discusses the advantages and disadvantages of old and new methods and provides recommendations for improved results from future inclinings.

1 Introduction

An inclining experiment is used to indirectly measure the vertical center of gravity of a ship by measuring resultant heel angles for a given weight moved athartships. The methods for measuring these angles are considered tried and true even though the uncertainties of their accuracy and precision are not well understood.

The purpose of this paper is to examine variations and uncertainties of both traditional measurement methods and modern electronic inclinometers for measuring heel angles. Although an inclining experiment has many factors that contribute to the overall uncertainties, this paper will only take a close look at the uncertainties connected with heel angle measurements.

It should be noted at the start that the underlying data used in the analyses in this paper are a mix of actual and simulated values. With the abbreviated data available, there was a need to estimate variation probability curves for some factors in the first part of this paper. However, the method of looking at sensitivity is useful even though the final results may not be fully indicative of the actual uncertainties of an inclining experiment.

Uncertainty analysis is a world of its own and an important one. I am not qualified to present the mathematics of uncertainty fully and I approached this paper with the idea of using simulations to explore sensitivities and to use some data analysis tools to explore the electronic inclinometer data. I believe the result will be usable in the real world even if it may not fit into the boundaries of a formal mathematical uncertainty analysis.

2 Traditional methods

In one way or another the methods used in an inclining experiment go back hundreds of years. They follow from the very practical difficulties of measuring the weight and center of gravity of very large objects. Fortunately a ship is a very large object that is floating which means the weight and center of gravity (CG) of it can be determined indirectly. That is, they can be determined from calculated factors—the volumetrics of the ship—along with empirical values from the inclining experiment. Thus an inclining experiment can have errors or uncertainties stemming from the calculation of volumetrics (hydrostatics) as well as from actual physical measurement errors. [1]

While this paper will not discuss possible errors from hydrostatic (volumetric) calculations, we have to acknowledge that the inclining itself is a bit of a hybrid or indirect measurement with all the issues that stem from that.

2.1 The Inclining Experiment

The following is a description of the inclining experiment taken, in part, from a previous paper of mine [Tel14].

The main purpose of the inclining experiment is to determine the lightship weight of the ship and its vertical center of gravity (VCG). The longitudinal center of gravity (LCG) is also calculated although we will not be discussing the LCG in this paper.

Details for performing an inclining experiment can be found in [CT07], [09616], and [Ame] but the basic procedure is to move a heavy weight athwartships (port and starboard) several times and measure the resulting angle of heel for each movement at different stations in the ship. This provides the measurement of the transverse stability in the condition of the ship at the time of the inclining. This measurement is the distance from the center of gravity (G) to the transverse metacenter (M): GM.

For each weight movement, the GM can be calculated as follows:

$$GM = \frac{wd}{W \tan\theta} \tag{1}$$

where:

$w =$ weight of the inclining weight

$d =$ distance the weight is moved athwartships

$W =$ displacement of the ship at the time of the inclining

$\theta =$ angle of heel

After several weight movements and different distances (thus different transverse moments), a plot is developed of the athwartships moment versus the tangent of the heel angle (see Figure 1). The GM for all the weight movements is calculated from the slope of the line fitted to all the data divided by the displacement:

$$GM = \frac{slope}{W} \tag{2}$$

This GM is compared to the calculated volumetric properties of the boat to derive the vertical center of gravity:

$$KG = KM - GM \tag{3}$$

The KM term is the distance from the keel to the transverse metacenter. GM is the distance from the center of gravity to the metacenter. Subtracting the two provides the distance from the keel to the center of gravity. [2]

[1] A more detailed discussion of volumetrics and hydrostatic calculations can be found in [Tel17].

[2] For submarines the baseline (reference plane for vertical measurement) is the keel so KG=VCG. For surface ships the keel often extends below the baseline so the KG and VCG are not equivalent. Throughout this paper we will use the term KG for VCG to make things simpler.

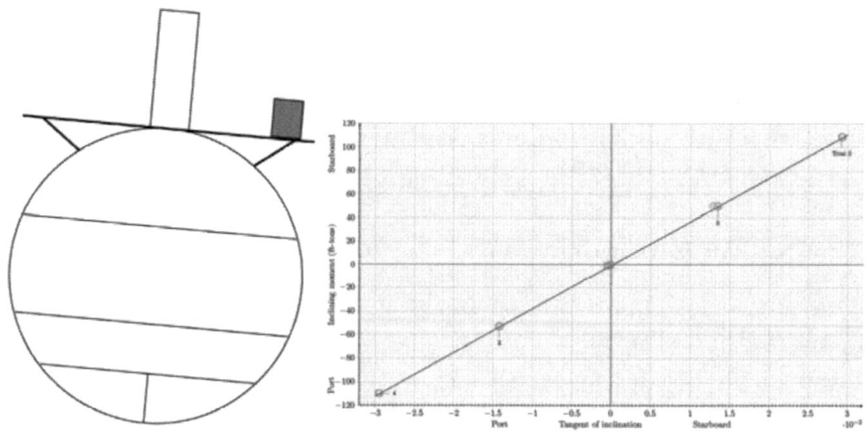

Figure 1: Inclining experiment setup for a submarine and resulting plot of tangents. The method is very similar for surface ships.

The weight movements and the resulting angles of heel provide the data to determine the KG at the time of the inclining (called the as-inclined condition). The inclining experiment also requires a complete inventory of the boat so that the as-inclined condition can be corrected back to a light ship baseline condition. Standard loads are added to the light ship to arrive at the normal operating condition. For our purposes we will not go into loads but will instead concentrate on how uncertainties affect the GM in the as-inclined condition. It is just arithmetic to extend this to the KG in other ship loading conditions.

One important thing to note in this general inclining procedure is the use of the Plot of Tangents to determine the GM. In itself this is a method to reduce uncertainty; by using multiple weight movements as well as multiple angle measurement stations the we can identify and reduce the impact of large errors. Any outliers of angle measurements can be identified in real time and corrections can be made immediately. That being said, it still means that there are usually only six to eight weight movements and three angle measurement stations so we normally deal with only about 18 to 24 data points.

2.2 Traditional tools

The traditional tools for measuring the angle of heel are old and simple. The oldest tool and the one that is still often used is the pendulum. A schematic of the setup is shown in Figure 2.

As you can see, the pendulum setup requires quite a lot of assembly and requires a space that allows for as long a pendulum length as possible, preferably 12 to 15 feet or more. Three of these stations must be placed along the length of the ship and as close to centerline as possible. They also need to be out of way as any disturbance by a passing crewmember may require restarting and recalibration of the station. So even in a large ship the number of available locations is limited.

The readings taken with the pendulum are the distances of the swing of the pendulum marked on a paper tape attached to the batten. Of course the ship is floating and moves with water movement, wind, personnel movement (though they

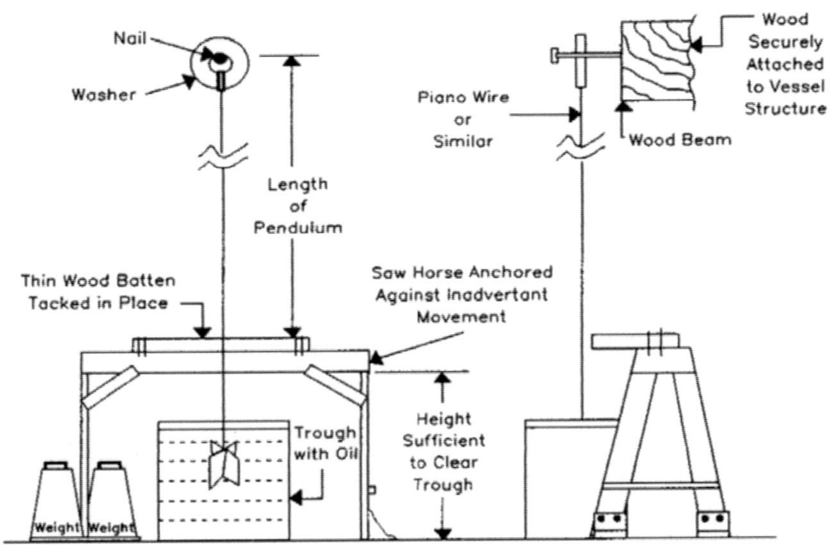

Figure 2: The rather involved pendulum setup.

are supposed to stand fast), and residual movement of fluids in tanks (this should be minimized by filling up or pumping down). The trough of oil (now usually dish soap) damps down the movement but the pendulum still swings a bit so the reader must estimate the upper and lower extents of the swing and report out the probable middle position.

From the measurement of the swing and the known length of the pendulum line the tangent of the heel angle is calculated. On the plot of tangents the tangent and the associated transverse moment are plotted for each reporting station. As the experiment continues, the plotter examines a potential best fit line for real-time indication of errors. Large variations in the reporting stations are also noted.

Another tool used is the U-tube as shown in Figure 3. This is essentially a water level apparatus that is installed on the deck. Usually three stations are set up along the length of the ship. Instead of the swing of a pendulum, the change in water level is observed at one end. Like the pendulum setup, the movements are marked on a batten and the operator must estimate the central measurement from the high and low extents of the water movement due to ship roll. The advantage to U-tubes is the increased length of the "pendulum", the easier setup and takedown, and the availability of more locations.

The final traditional tool that is still in use is the mechanical inclinometer like that shown in Figure 4. The inclinometer is used on ships and submarines where there isn't height or space for pendulums or U-tubes. These devices are very precise, very expensive, and require a good bit of training to get consistent and good readings from them.

The inclinometer reads the tangent of the heel angle directly and usually has

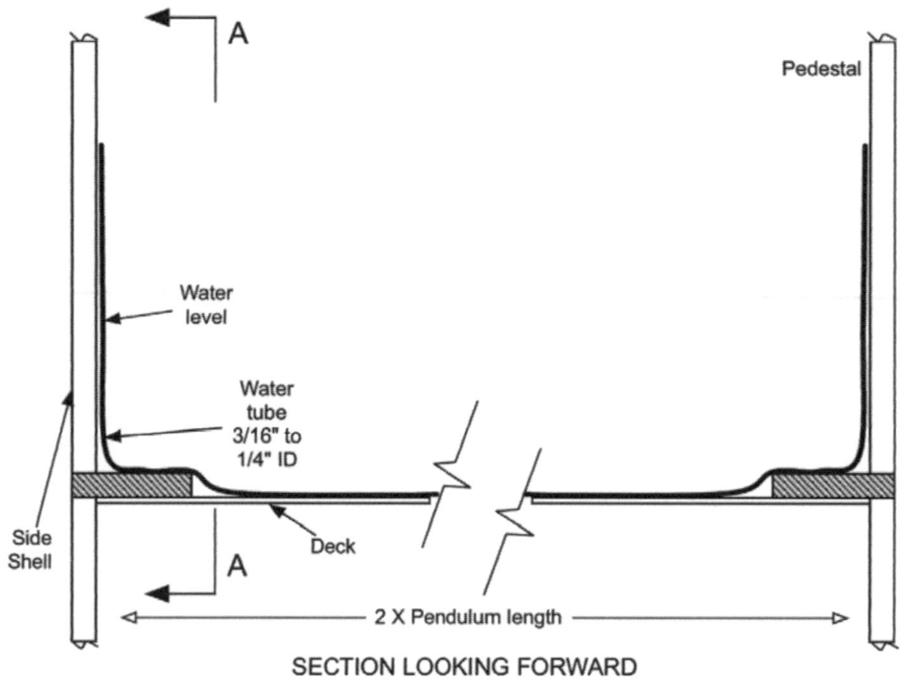

Figure 3: The U-tube setup.

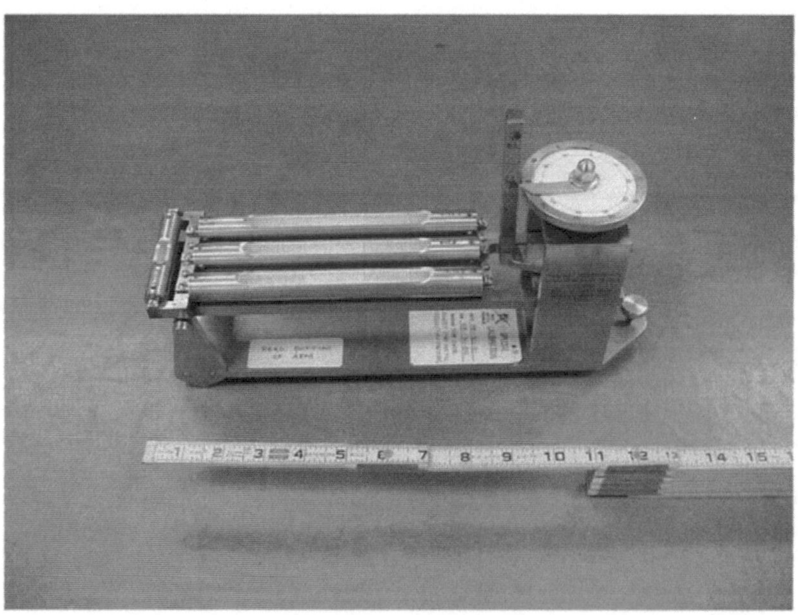

Figure 4: The mechanical inclinometer. This one has three bubble levels of varying curvature or "speed". The tangent of the heel angle is read directly from the vertical scale and the disk.

two or three bubble levels of different curvatures or "speeds". Just like the other methods, the reader must interpret the movement of the bubbles and estimate the middle point of rolls. This usually requires experience and training to get it right.

The advantage of the mechanical inclinometer is the easy setup, its portability, and its precision. The direct tangent readings are usually to the ten thousands place (0.00001).

3 Electronic Inclinometers

Electronic inclinometers have been used experimentally in inclinings for the past thirty plus years and yet have never been fully approved by either the US Navy or the US Coast Guard. As of recently (to the best of my knowledge) the ASTM guide that the US Coast Guard uses [Ame] stated the following:

> "If the person conducting the test desires to substitute inclinometers or other measuring devices for the pendulums, complete prior testing of the measuring devices to verify their accuracy before actual substitution for the pendulums. It is recommended that the inclinometers or other measuring devices be used in conjunction with the pendulums instead of using only other devices and no pendulums."

This reticence to use other angle measurement devices has been tough to combat. The older methods are considered tried and true and can produce the data that are accurate and precise enough for most purposes. But technology has distinct advantages of portability, ease of setup, expense, accuracy, precision, and most importantly the ability to measure and record continuously thereby providing many more data points that can be analyzed later.

Figure 5 shows two electronic inclinometers. The top one is an early model built by the author some thirteen years ago. This was a proof of concept device that was never used in a real inclining experiment. It used a commercially available accelerometer and an analog to digital converter and was meant to provide real-time data during an experiment. This unit was small, cheap, and accurate enough for most inclinings (in the author's opinion).

The lower unit is a commercially available logging inclinometer. This was used for two inclining experiments, the data for one is used in the next section of this paper. This inclinometer costs less than a hundred dollars and runs for long periods on a single AA battery. It logs movement in three axes at 10 Hertz or greater.

Naturally the disadvantage of this device is that the data is not seen in real time. But this is an old unit and these are now no doubt available with WiFi or bluetooth capabilities.

Though I have not used one in an inclining, it is likely that the accelerometers in most smart phones today would be accurate and precise enough to measure inclining heel angles.

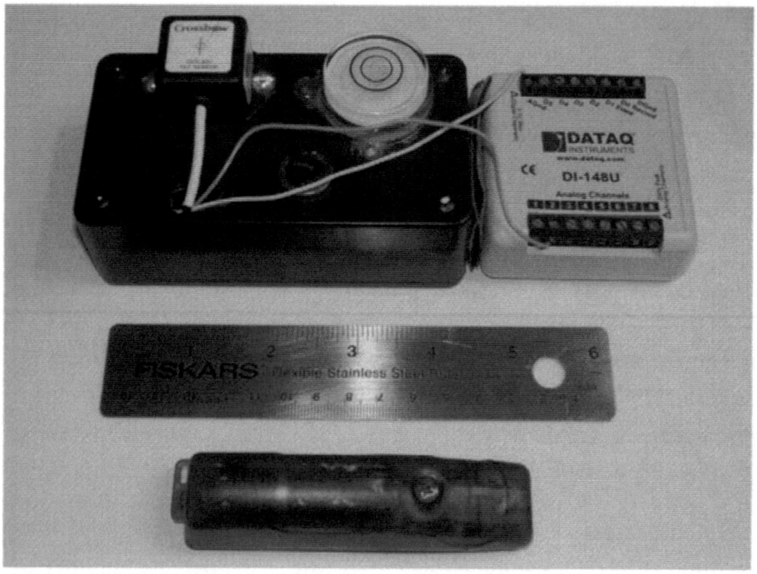

Figure 5: At top is a homemade 2-Axis electronic inclinometer with a commercial sensor and analog to digital converter. The lower unit is a logging 3-axis inclinometer.

4 Sensitivity analysis — traditional

As mentioned before, a full uncertainty analysis was not done for this paper[3]. Instead, for the traditional measurement methods, I looked at the sensitivity of the system to changes (or errors) U-tube lengths and in the measurements of deflections of the U-tube water levels for each inclining movement.

4.1 The tool

Instead of just changing the values for U-tube lengths and deflections and looking at the resultant change to slope of tangents and as-inclined stability (GM), I wanted a tool similar to the one used in [Tel05]. Unfortunately the software tool used in that paper and in some subsequent papers was not available for this study. Instead I programmed a simulation tool in R [R D08] which is capable of introducing errors to all of the inclining calculation factors including: water density; forward draft reading; aft draft reading; the weight of each inclining weight; the distance each inclining weight moved; the weight, vertical moment, and longitudinal moment of weights to complete, weights to relocate, and weights to deduct; the free surface moment; the length of each pendulum or U-tube; the deflection of each pendulum or U-tube level; and the inclinometer readings.

This tool allows the user to pick three different probability distributions for each factor: uniform, triangular, and normal. Thus the results, in this case the GM as-inclined, can be shown as probability distributions and one can make some conclusions of error ranges and uncertainty based on the usual tools applied to probabilities.

[3]Although it may be revisited for a future paper.

4.2 Nominal condition

For this paper, 2000 points were used for each probability distribution and only the lengths of the U-tubes and their deflections were varied. All other factors were kept at the nominal, reported value. The inclining used was for a very old surface warship using three U-tube stations and five weight movements: half way to starboard, all the way to starboard, back to centerline then halfway to port then all the way to port and then back to centerline. In this experiment the first movement was repeated after the full starboard movement and this actual third movement was used as the first movement.

For each of the 2000 runs a value was taken from each of the probability distriburtions and added to the nominal values for each U-tube deflection or length measurement. Thus for each movement we have 6000 data points encompassing the full range of variation applied to the nominal measurement value.

The plot of tangents for the baseline or nominal condition (no variations) is shown in Figure 6. Note that the slope calculated by the tool (9945) is slightly different than the one in the written report (9956). To be consistent we will use the 9945 for comparison purposes.

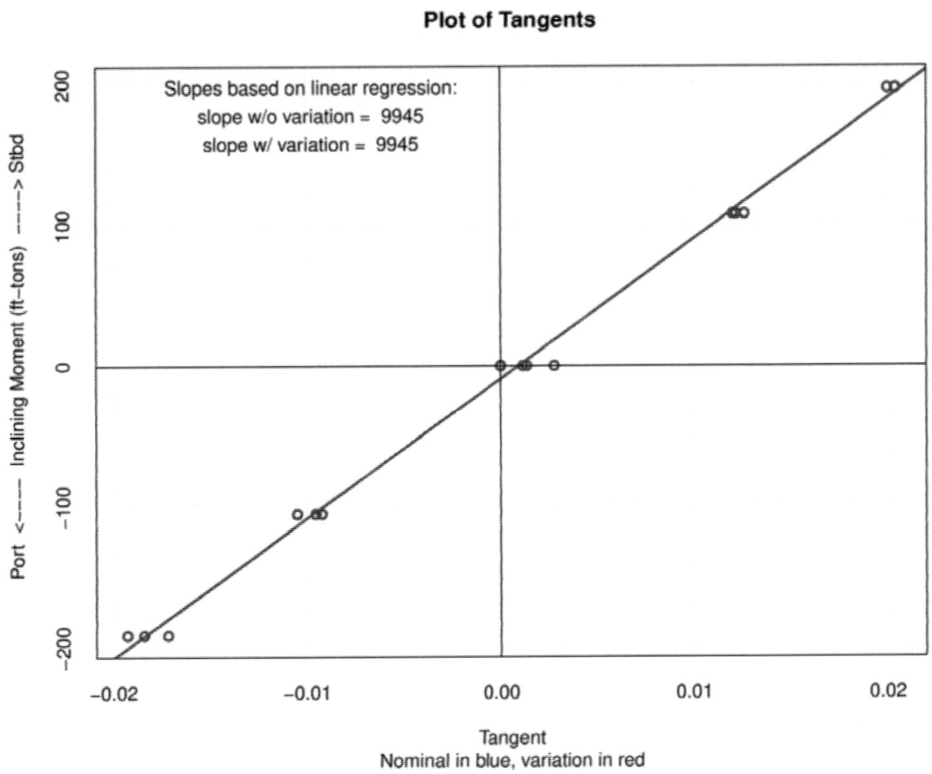

Figure 6: Plot of tangents for the nominal or baseline condition without any variations.

The as-inclined displacement for the nominal condition is 1965 long tons so:

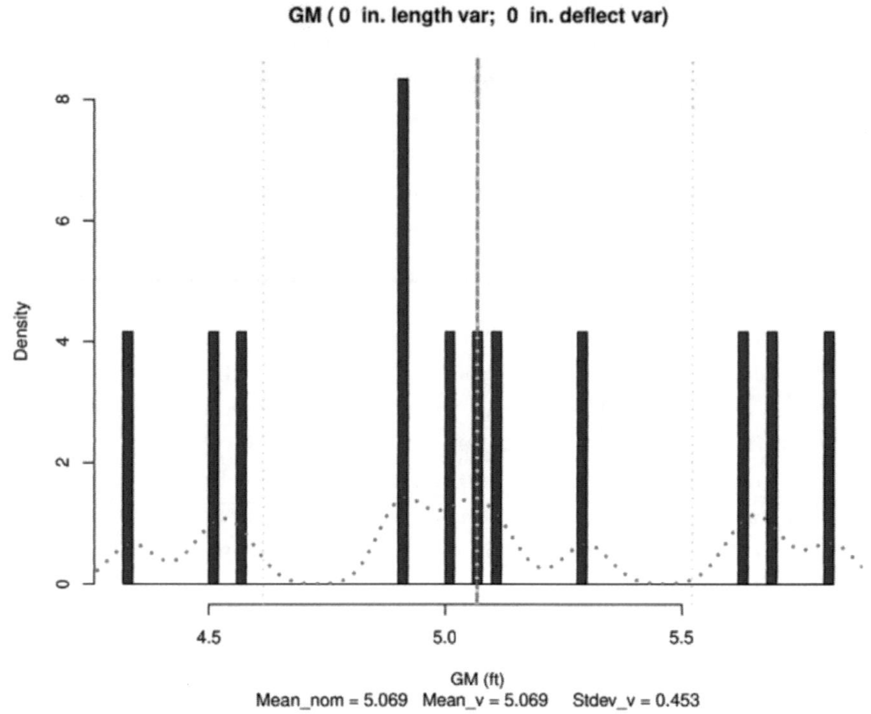

Figure 7: Distribution of GMs from the nominal condition.

$$GM = \frac{slope}{W} \tag{4}$$

$$GM = \frac{9945}{1965} \tag{5}$$

$$GM = 5.06 \text{ ft.} \tag{6}$$

If we look at the GMs calculated from the individual stations and movements as a distribution (see Figure 7), we find that the range of GMs is from 4.34 ft to 5.82 ft with a mean of 5.07. In this nominal case the linear regression results in a GM value close to the mean of the distribution of values and close to the manual regression value of 5.07 that was stated in the inclining report.

4.3 Variations added

With the nominal condition and values known, we can start introducing variation into the U-tube deflections and into the lengths of the U-tubes to understand how those two factors influence the as-inclined GM.

Length var.	Deflect var.	Min	1st Q	Median	Mean	3rd Q	Max	Slope
0	0	4.34	4.821	5.035	5.069	5.38	5.815	9945
0	0.03125	4.25	4.806	5.035	5.069	5.409	5.998	9944
0	0.0625	4.183	4.784	5.025	5.07	5.409	6.187	9943
0	0.125	3.997	4.775	5.02	5.072	5.413	6.507	9934
0	0.25	3.628	4.729	5.027	5.075	5.419	6.748	9908
0	0.5	3.15	4.62	5.036	5.098	5.518	7.613	9800
0	1	2.477	4.393	4.973	5.083	5.656	8.736	9383
0.0625	0	4.338	4.82	5.035	5.069	5.383	5.819	9945
0.125	0	4.335	4.818	5.036	5.069	5.385	5.822	9945
0.25	0	4.331	4.815	5.036	5.069	5.388	5.828	9945
0.5	0	4.322	4.81	5.038	5.068	5.397	5.841	9944
1	0	4.297	4.792	5.037	5.068	5.406	5.862	9944
3	0	4.209	4.742	5.033	5.068	5.457	5.981	9941
1	0.25	3.485	4.731	5.025	5.076	5.421	6.74	9908
1	0.5	2.953	4.621	5.032	5.096	5.513	7.583	9793
1	1	2.259	4.39	4.989	5.14	5.695	10.1	9381
3	1	2.33	4.393	4.983	5.122	5.705	9.25	9403

Table 1: Summary of GM from runs with variations added to U-tube deflections and lengths.

4.3.1 Variation of deflections

For examining the influence of the deflections of the U-tube levels, we chose a variation normal distribution with a mean of 0 and standard deviations of 1/32 of an inch to 1 inch for all three U-tube stations. This distribution of variation was added to the nominal values for all three U-tube stations.

The summary of results of GM is shown in the top portion of Table 1.

Figure 8 shows the distribution of GMs for the case where the variation in each deflection was a normal distribution with a mean of 0 and a standard deviation of ± 1/8 inch. The 1/8 inch is a normal expected accuracy of measuring both the length of the U-tube and the deflections with a goal of 1/16 inch.

The slopes of tangents show little change until the 1/2 inch variation is introduced whereupon the resultant slope is 9800 (compared to the nominal 9945). This would put the as-inclined GM at 4.99 instead of the reported value of 5.07. This suggests that even though the increased variation results in GM distributions of increasing range, the linear regression method for the slope of tangents is relatively resistant to these errors in deflection measurement. It should be noted that with all these simulated data points the slope of tangents is not necessary to determine a GM value, but it is illustrative of the effect of introducing variations into the calculations.

Taking to some extreme point of mis-measurement, Figure 9 on page 13 shows the distribution of GMs with a variation on the deflection measurement of the U-tubes of one inch. The resultant plot of tangents in Figure 10 shows that now the variation has shifted the best fit line (red dotted line) and the slope is now 9383

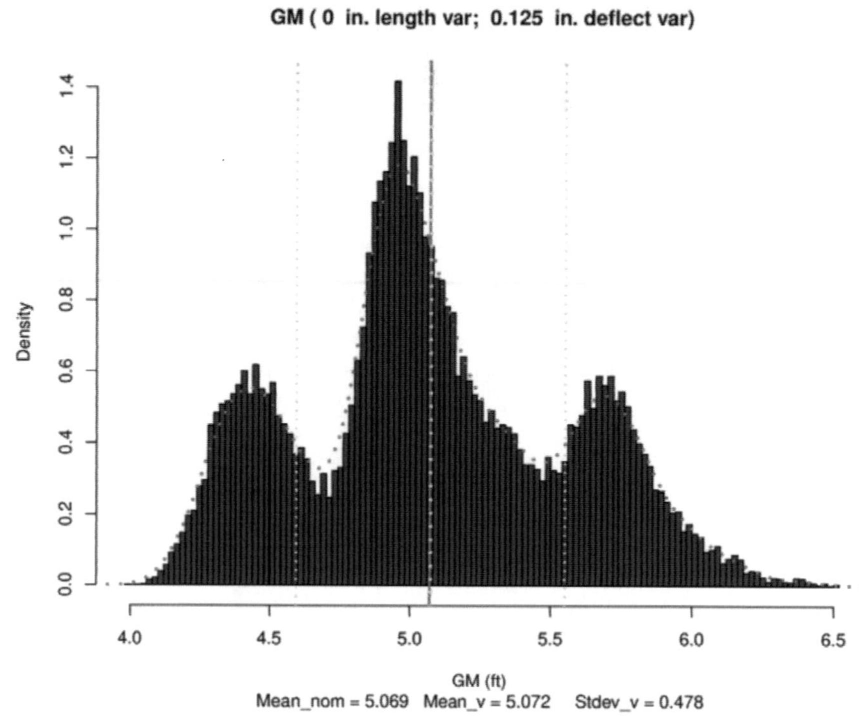

Figure 8: Distribution of GMs from the condition with a variation to U-tube deflections of 1/8 inch standard deviation.

which would mean a calculated GM of 4.78 ft.

4.3.2 Variation of lengths

Since the U-tube lengths are in tens of feet, one would expect similar measurement variations applied would result in less variation in resultant GMs. The middle part of Table 1 shows that this is the case. The maximum length variation in that table is three inches which, in real life, is a ridiculously unlikely error. Still, it shows that the mean of the GM values doesn't change much (as expected since we apply the variation equally plus and minus) and the range of variation is not as large as with the variation of the deflections.

As one can see in Figures 11 and 12 on page 14, for a one inch variation in the lengths of the U-tubes, the distributions of GM are still around the individual U-tube readings and the slope of tangents is essentially the same as the nominal condition.

From this we can assume that the resultant GM as-inclined is relatively resistant to errors in the measurement of the U-tube lengths.

4.3.3 Variation applied to both

Finally, we can look at the results of adding variation to both the length and the deflections of the U-tubes. The values are shown in the bottom of Table 1. Figures

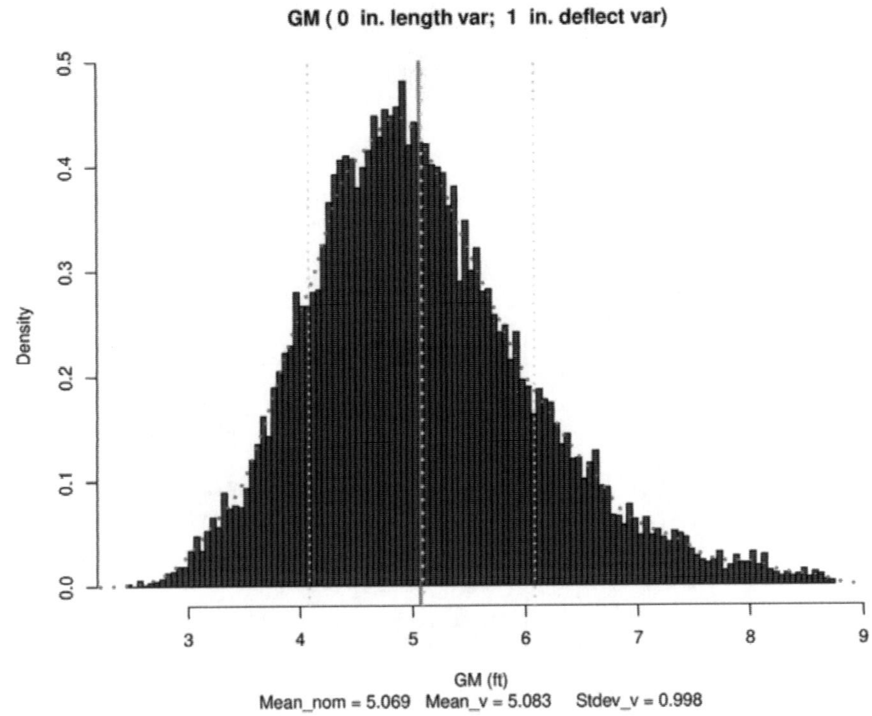

Figure 9: Distribution of GMs from the condition with a variation to U-tube deflection of 1 inch standard deviation.

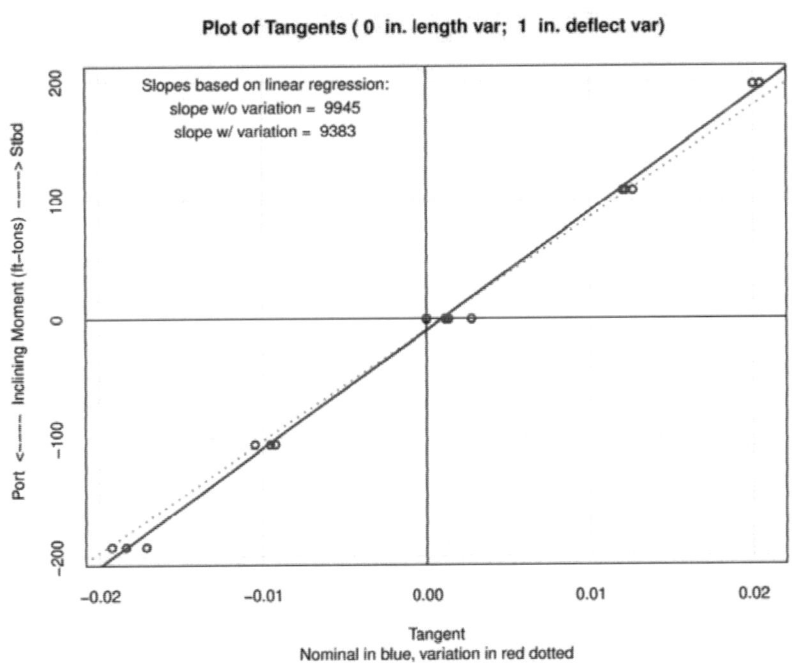

Figure 10: Plot of tangents from the condition with a variation to U-tube deflection of 1 inch standard deviation.

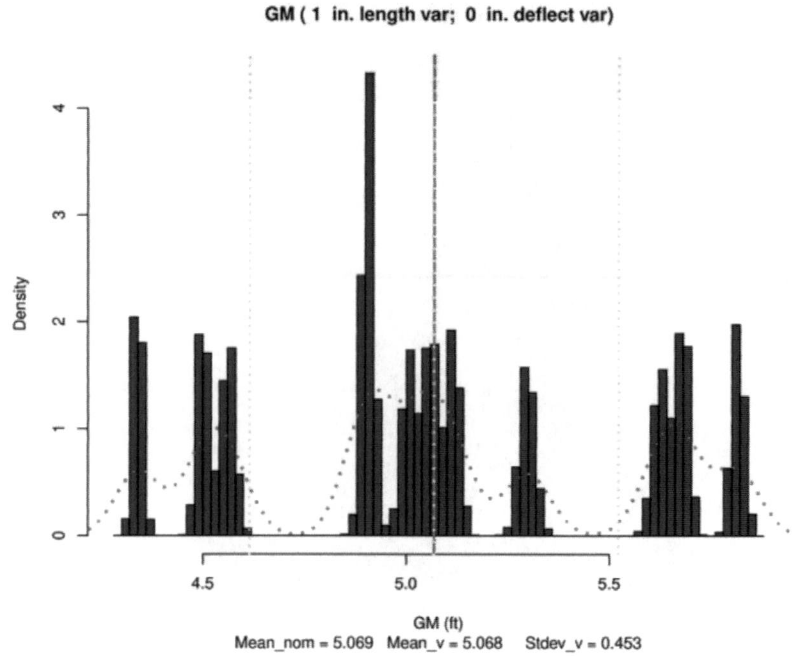

Figure 11: Distribution of GMs from the condition with a variation to U-tube length of 1 inch standard deviation.

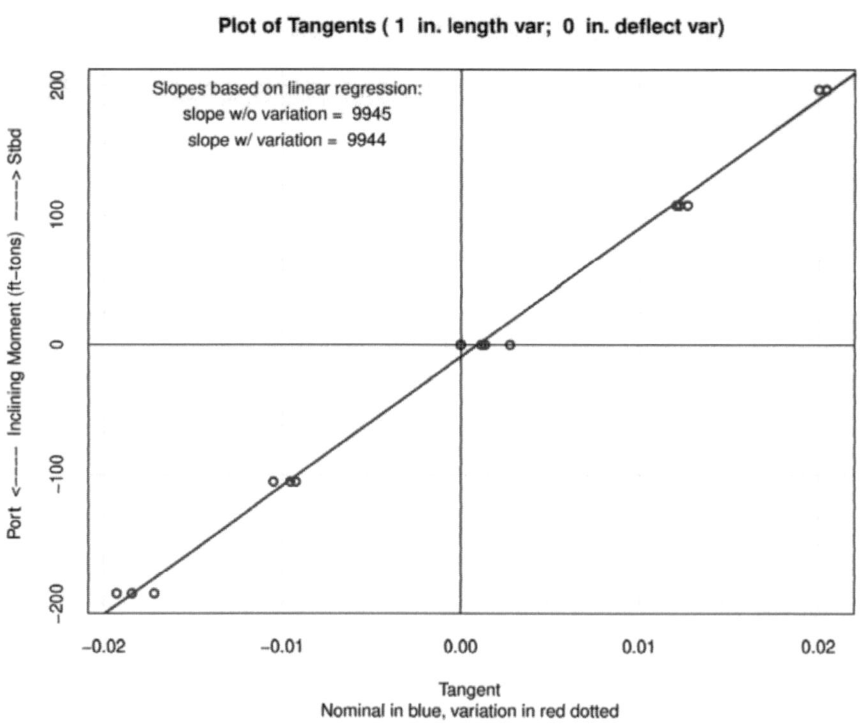

Figure 12: Plot of tangents from the condition with a variation to U-tube length of 1 inch standard deviation.

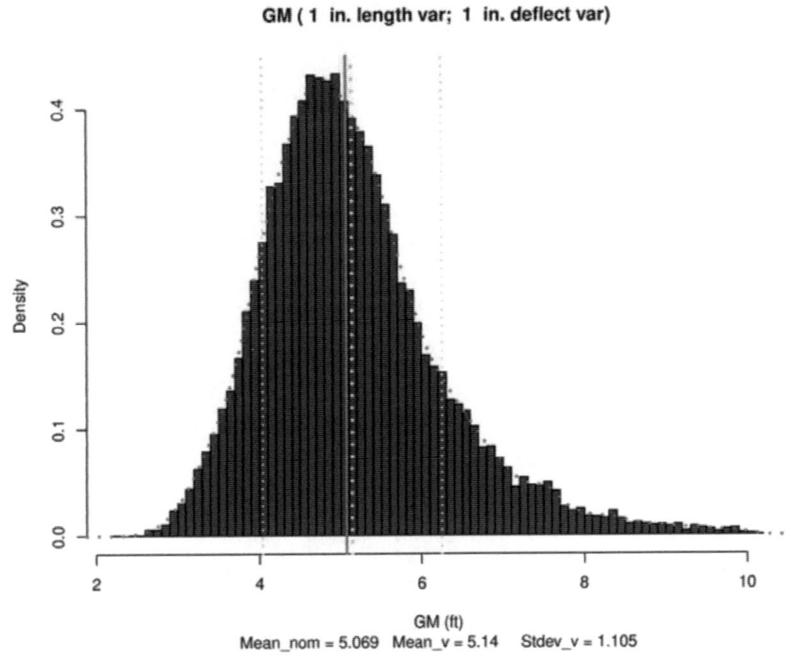

Figure 13: Distribution of GMs from the condition with a variation to U-tube length and deflection of 1 inch standard deviation.

13 and 14 on page 15 show the distribution and slope of tangents for an extreme case of a one inch variation applied to both the deflection and the measured length of the U-tubes.

4.4 Observations

While this was by no means a full uncertainty study on the effect of errors of U-tube deflection and length measurements, it does indicate two things: 1) that the slope of tangents method is still reasonable and somewhat robust, and 2) that applying some normal distribution of measurement errors (with no positive or negative bias) will increase the range of results but the mean values will be quite similar to the nominal values in the submitted report.

This second point identifies the issue with the tool used and with any other attempt to quantify inclining experiment errors. Which is that there is no way of knowing if these errors should be biased in one direction or the other. The tool does, however, provide some insight into the range of GMs as-inclined given possible measurement errors and would allow us to report the GM as, say, 5.072 ± 0.478 ft (the mean and one standard deviation for the 1/8 inch deflection calculation). Of course this would allow some parties to immediately grab onto the larger number and run with it. It should be noted that the traditional inclining experiment variation used for KG is 0.25 ft. [07916]

While these traditional measurement methods may seem archaic, they do produce valid values if done correctly. The use of three stations, multiple weight move-

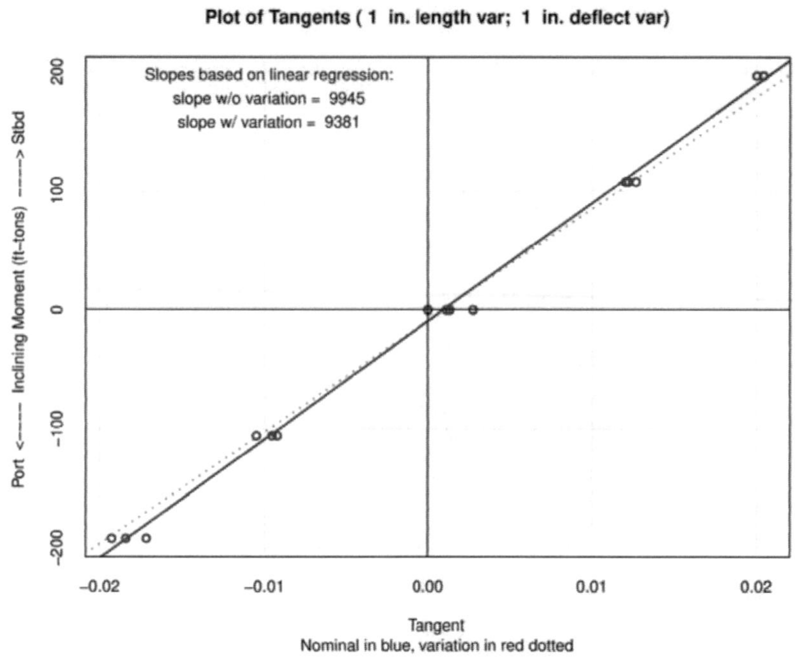

Figure 14: Plot of tangents from the condition with a variation to U-tube length and deflection of 1 inch standard deviation.

ments, and a linear regression of the data very much reduces the impact of any measurement errors.

Finally, this little exercise corroborates previous sensitivity studies [Tel05] which show that other factors in the inclining experiment have greater impact on the results and thus the errors for those factors (like weight and vertical moment of weight to deduct) should be minimized before concentrating on deflection measurements.

5 A modern method

While traditional methods can still produce good results, technology has moved on and there are ways to get similar or better results with cheaper, smaller, and more precise equipment that is also faster and easier to set up.

Here we will look at the same inclining experiment on the same ship but with the data provided by the little blue electronic logging inclinometer previously shown in Figure 5 on page 8. It should be noted that if the electronic inclinometer were to be used as the primary source of data for the inclining, two more units would have been used, spaced longitudinally along the ship. This would have provided some 320,000 more data points to examine than the data we have here.

5.1 Raw data

The raw data from the inclinometer is shown in Figure 15. This is a timeline of the entire inclining experiment with a heel angle reading (in tangents) made every 1/10

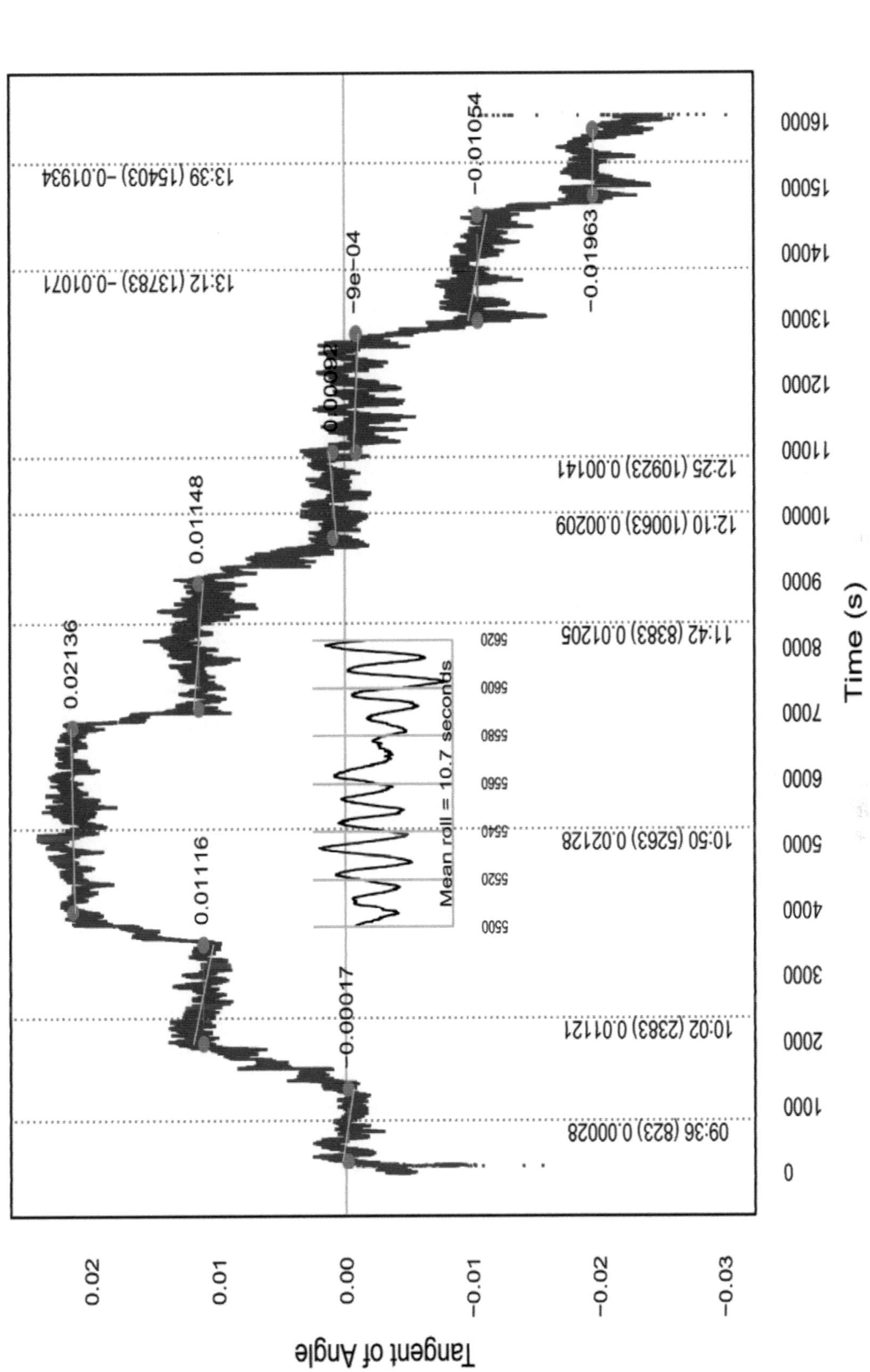

Figure 15: A timeline of the inclining experiment showing the tangents of the angles for all weight movements. The red dotted lines show the time when the U-tube readings were taken. The local time, the seconds from the start, and the reported tangent value is shown in the vertical labels. The red lines connected by red dots indicate the mean tangent from all the data at that weight movement. The green line shows a best linear fit for that movement.

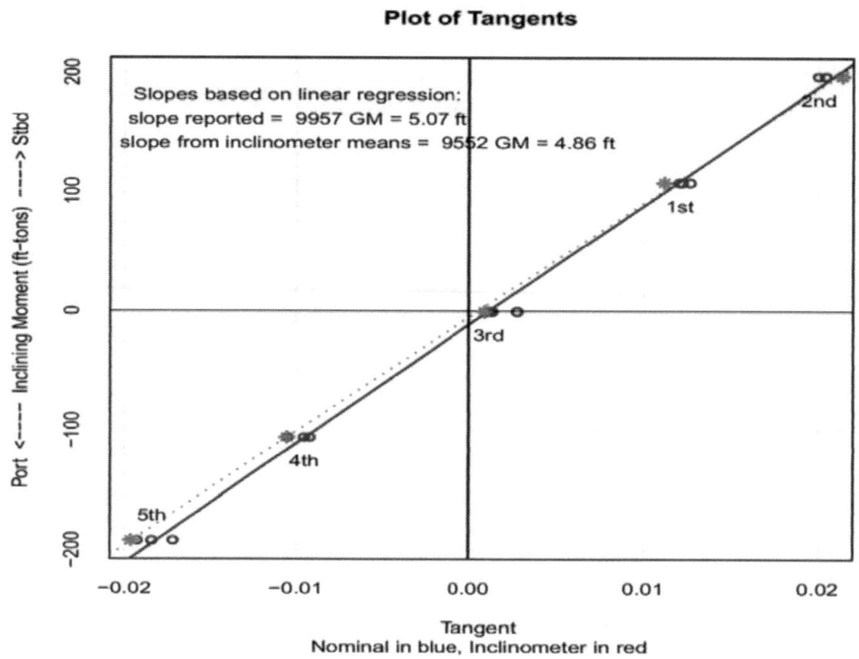

Figure 16: Plot of tangents showing the reported values and line (in blue) and the overlying plot of points of the mean values for each movement from the electronic inclinometer.

of a second (10 Hz). This results in 166793 readings covering the entire 4 hour and 38 minute experiment. The data were corrected for initial zero point (no ship heel) and for local time.

The red dotted lines show the time when the U-tube readings were taken. The local time, the seconds from the start, and the reported tangent value is shown in the vertical labels. The red lines connected by red dots indicate the mean tangent from all the data at that weight movement. The green line is a best linear fit for that movement.

5.2 Data analysis

Figure 16 shows the original plot of tangents along with the plot of the means from the electronic inclinometer. Of course one of the advantages of the inclinometer is that we have much more data for each movement and we are not restricted to one reading from each station per movement or even one mean value from each station. We can look at the data from each movement and make a determination of the best value or values to use. We can also make more general deductions from the data.

For instance, let's take a closer look at the first movement. Here the reported tangents were 0.01202, 0.01264, and 0.01219 for a mean of 0.01228 and an instantaneous GM of:

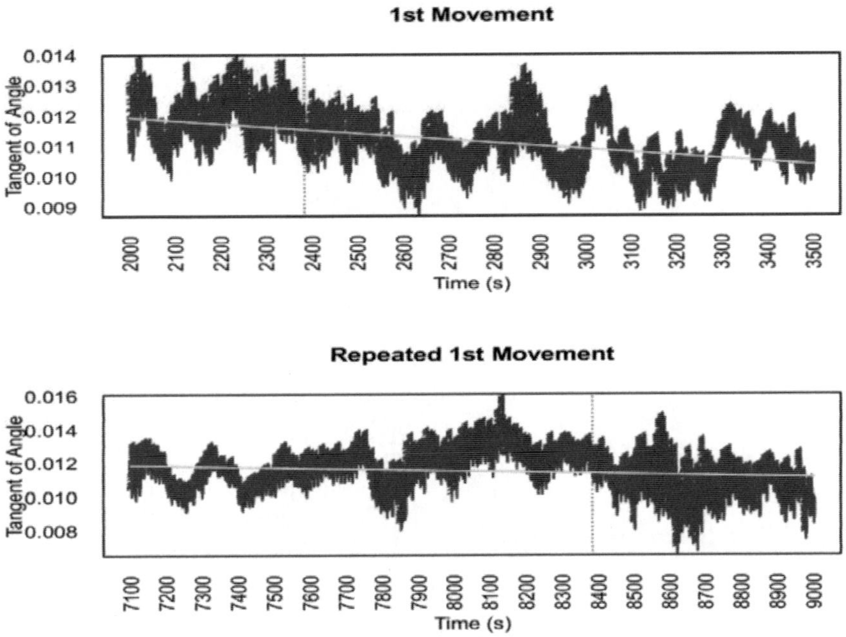

Figure 17: Raw data for the first weight movement and the repeated first movement. The times of the U-tube readings are shown with a dotted red line.

$$GM = \frac{wd}{W \tan\theta} \tag{7}$$

$$GM = \frac{107.8}{(1965 * 0.01228)} \tag{8}$$

$$GM = 4.467 \text{ ft} \tag{9}$$

However, the first movement was repeated for an unreported reason. So the inclining continued with the second movement to starboard and then a second first reading after that. This can be seen in Figure 15 as the movement at about the 8300 time mark.

In the past we could only trust the experience of the team to choose to repeat a movement and use what they felt was the better data, but with the electronic inclinometer we can look and see what the issue might have been and we can make a more informed decision about the data to use.

So the initial first movement is from the 2000 time mark to the 3500 time mark which equates to 15000 data points.

Figure 17 shows the raw data for the first movement and the repeated first movement. The vertical red dotted line shows the approximate time that the U-tube readers recorded their values. The inclinometer showed a value of 0.01252 and 0.01405 respectively for these times. The instantaneous GM for the first movement

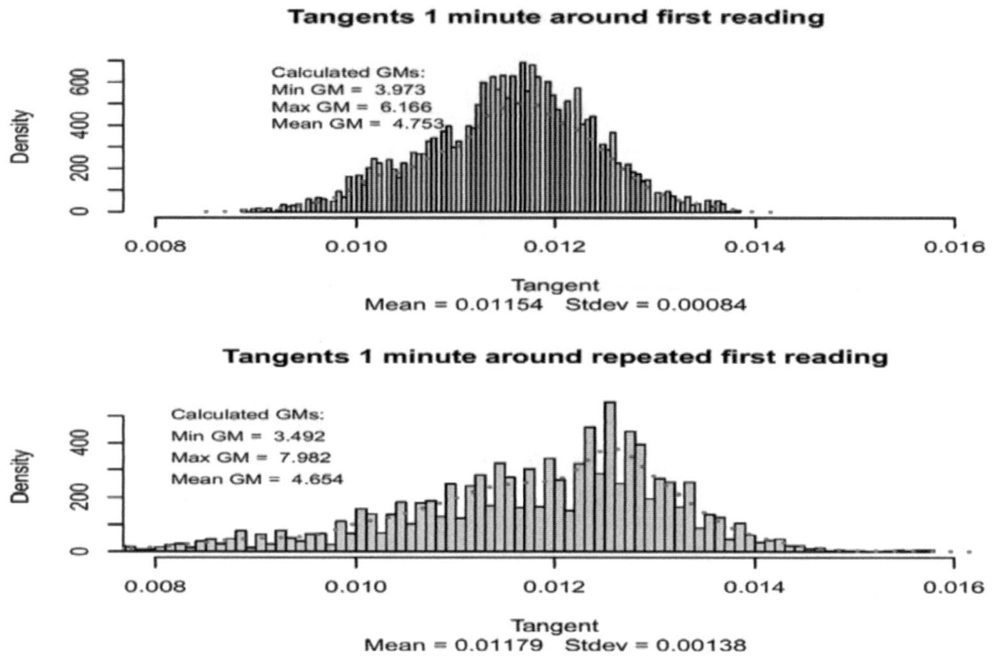

Figure 18: Histograms of the data from the electronic inclinometer for the one minute around the time that the U-tube readings were taken for both first movements.

that was not used would be:[4]

$$GM = \frac{107.8}{(1965 * 0.01252)} \tag{10}$$

$$GM = 4.382 \text{ ft} \tag{11}$$

And for the repeated first movement:

$$GM = \frac{107.8}{(1965 * 0.01405)} \tag{12}$$

$$GM = 3.904 \text{ ft} \tag{13}$$

But with all that data at hand, and not knowing exactly when the U-tube readings were recorded, lets take the data from a minute around the reported recording time.

Figure 18 shows the histogram of the tangent readings from 30 seconds before and after the reported time of the U-tube readings for the first movement and for the repeated first movement. One can see a large range of tangents and the calculated

[4]The GM values assume that the inclining moment was the same for both first movements; the report only contains the moment for the repeated movement.

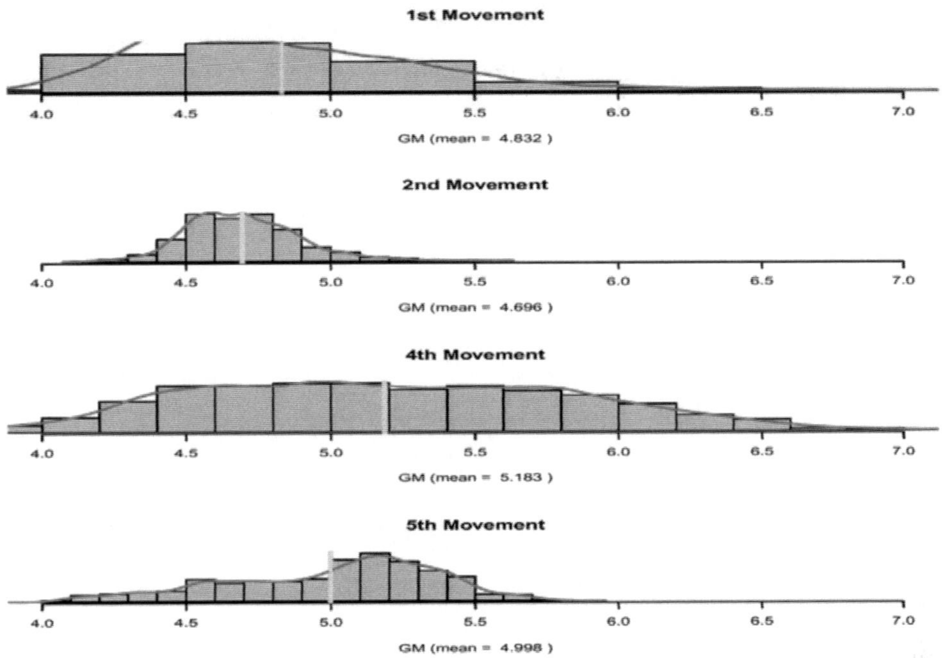

Figure 19: Histograms of the data from the electronic inclinometer for each of the inclining movements.

GMs from minimums of 3.973 and 3.492 ft to maximums of 6.166 and 7.982 ft for the first and repeated first movement respectively.

Of course this illustrates exactly why we don't use the instantaneous GM in the traditional methods but it also illustrates the importance of skilled people reading the U-tubes. By carefully watching the rise and fall of the U-tube water level, the reader estimates the mean value. In this case the calculated GM from the mean of the reported tangents was 4.467 ft which is credibly close to the mean of 4.654 ft from the electronic inclinometer readings shown in Figure 18. But the range of tangents shone for just a minute of duration shows how sensitive this measurement is. Naturally it follows that using the electronic inclinometer means we don't need to have highly trained readers at each station and that we can make appropriate decisions on what data to use later on.

Looking at Figure 17 can also inform us on some more general concerns. It is clear that something was affecting the ship later on in the first movement, as the tangents decreased as the green best fit line shows. By inspection one can see that something influenced the ship starting around the 2500 second mark. By having this data available, we can go back and see if there was an environmental change at that time—wind, waves, currents—or if there was excess crew movement or tightening of mooring lines. Whatever that might have been, using traditional methods would not allow us a view of the data for the entire period and how those data changed.

This can lead us to compare the distributions of readings for each movement to see how ship motions affected it and to make some judgement for the quality of readings relative to each other. Figure 19 shows all four movements (not including

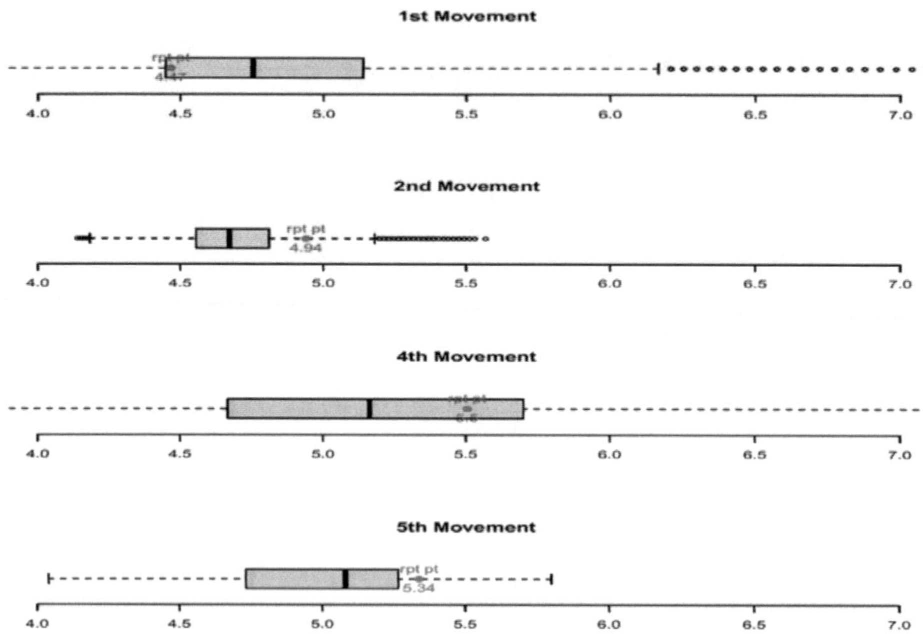

Figure 20: Boxplots of the data from the electronic inclinometer for each of the inclining movements.

Move	Min	1st Qu	Median	Mean	3rd Qu	Max	St Dev
1	3.492	4.448	4.756	4.832	5.138	7.982	0.53
1	4.139	4.556	4.674	4.696	4.812	5.569	0.2
1	3.376	4.668	5.163	5.183	5.699	7.376	0.67
1	4.039	4.733	5.08	4.998	5.264	5.8	0.36

Table 2: Summary of inclinometer data for each movement

centerline movements) and the distribution of calculated GMs. While the means are similar it is clear that the fourth movement (first movement to Port) shows a lot more range and might be worthy of a closer look to see what might have happened during that part of the experiment.

Figure 20 shows the same movements with boxplots showing the minimums, maximums, first and third quartiles, and the mean of all the data for each movement. The reported value from the U-tube reading is shown in red. Note that the reported value does not fall within the box except for the fourth movement (likely because it has such a large range). Table 2 shows the numerical summary of these data.

Figure 21 shows the histogram and boxplot of all the movements. Here the reported GM from the slope of tangents, 5.07 ft, is within the mid-fifty percent range. However, both plots show the large number of readings which may be outliers. Figure 22 shows the figure after removal of those outliers.[5] In this case it has not

[5]All values that are greater than 75th percentile value + 1.5 times the inter quartile range or lesser than 25th percentile value - 1.5 times the inter quartile range, are tagged as outliers.

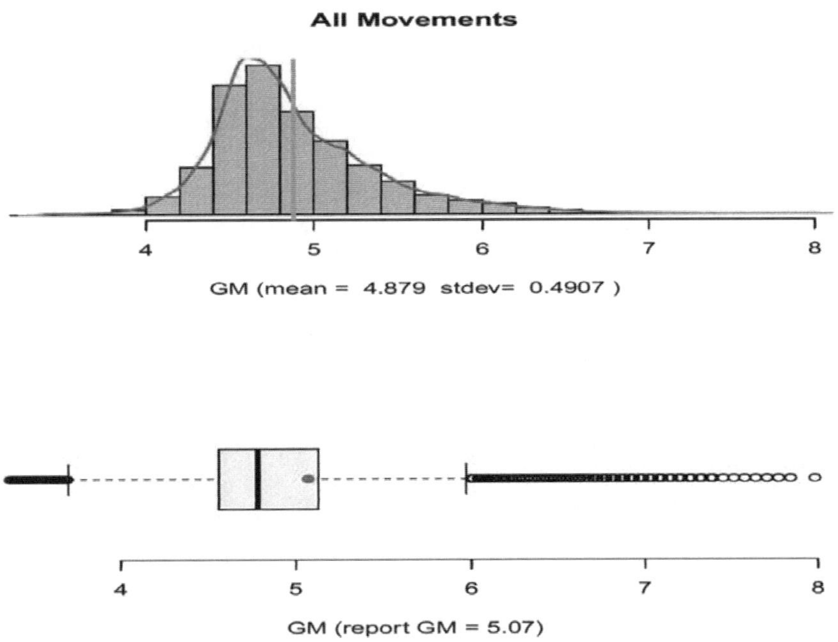

Figure 21: Histogram and boxplot of all the movements.

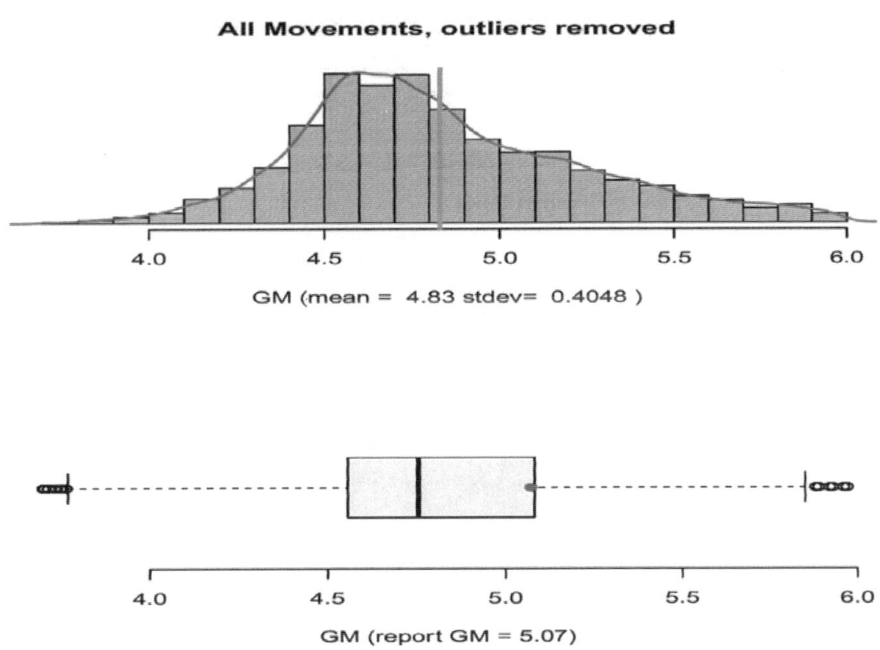

Figure 22: Histogram and boxplot of all the movements after removal of suspected outlying points.

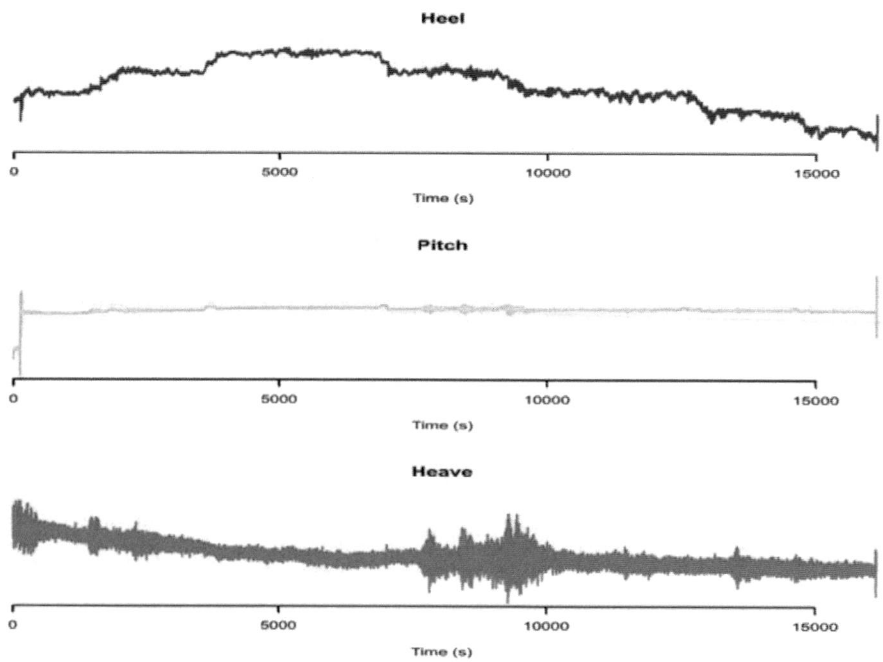

Figure 23: Plots of raw data for all three axes.

appeared to alter the answer much except to reduce the standard deviation. From this figure it appears that the result from the traditional method may show better stability than the ship actually has.

Naturally, by seeing all these data, one can make a more informed decision about what data to choose to use for the final stability calculations, or what data not to include.

Another apparent issue with the inclining that can easily be seen is at the return from starboard to centerline (around the 9800 second mark) which should have been a return to the initial condition with a zero moment and no heel. This was probably wind related as the entire slope line is shifted to starboard as can be seen in the Plot of Tangents in Figure 16. However, the data shows the ship was also affected between the 11000 and 12500 second marks which pushed it off the zero condition. This might have been transient or it might be an indication of crew or fluid shift that might have affected the following two weight movement. Without all these data this would not have been, and wasn't, seen. Whether it would make a huge difference to the final answer is hard to say, but at least having the data makes such an analysis possible, in combination with timed observations of environmental and ship conditions.

5.3 Bonus features

The electronic inclinometer also provides some bonus features that we have never had available for inclining experiments. The first is data on the other two axes: pitch and heave. Figure 23 shows the raw data plots of each axis. While we do not have enough information to make any quantitative judgements, it is clear that there was more ship movement in the 7500 to 10000 second time period and it might have been a factor in the data from the repeated first movement.

Finally, in looking at Figure 15 on page 17 you can see a small sub-plot of roll angle for the time marks of 5500 to 5620 or two minutes of roll data from the second weight movement. This data is used to calculate a mean roll period. Traditionally, this was measured by sallying the ship (inducing a large roll) and then timing three or more rolls and calculating a mean roll period. This data is used in comparing the sally constant with other ships of the class and with results from previous inclining experiments to give a rough estimate of GM from the roll period. Pulling this data out of the inclinometer readings makes the entire sally procedure unnecessary.

6 Conclusions

We have discussed the traditional methods of measuring the heel angle of a ship during an inclining experiment as well as a modern method. There are advantages and disadvantages to all the methods. But from the analysis of the data presented, it is clear that used properly all the methods are capable of producing credibly accurate results.

What sets the methods apart are the time, effort, and expense of setup/breakdown, the sources of possible error (human versus electronic), and the amount of data each method provides. This last factor is of great importance as there is only so much you can do with a limited data set. This paper attempted to show that some error can be somewhat artificially introduced to try to understand possible uncertainties, but the shape of those introduced errors may not correlate to the real world. And the errors are only applied to the physical measurements of U-tube deflection and length and (for this simulation) ignore any human errors in the actual readings.

The electronic inclinometer simply provides a massive amount of data for the entirety of the inclining experiment, not just discrete points at each weight movement. With the assumption that the instruments are properly calibrated, the data can be examined in minute detail (if required) to fully understand the range of uncertainty and to decrease those uncertainties by judicious selection of a credible subdata set.

6.1 Advantages and disadvantages

Pendulums, U-tubes, and mechanical inclinometers:

Advantages:

- Tried and true methods that are accepted by all certifying bodies.
- Geometry is simple to understand and the tangents easy to calculate.
- Some memory of the movements is retained in the marked up battens.

Disadvantages:

- Significant time and effort to set up and break down measuring apparatus.

- May require large vertical or horizontal spaces.
- Requires trained personnel for setup and reading. Prone to human error.
- It can be difficult to determine a good mean reading due to ship movement.
- Provides only three data points per weight movement or 15 to 24 data points total for a typical inclining.
- Apparatus is prone to disturbance by crew or weather. Tripping over a U-tube or mechanical inclinometer can put it out of use for the rest of the experiment.

Electronic inclinometers

Advantages:

- Almost no time or effort to install the apparatus.
- Requires very little deck or bulkhead space.
- Required few personnel, trained or otherwise, to set up.
- Provides continuous data throughout the experiment, in this case over 160,000 data points.
- Provides insight into adverse ship motions.
- Easier to protect the instruments from crew and weather influences.

Disadvantages:

- Accuracy and precision must be proved for certifying bodies; method is not universally accepted.
- Requires active electronic connection to see data in real time.
- Requires proper calibration or understanding of measured corrections needed.

6.2 Recommendations

I believe the advantages of the electronic inclinometer far outweigh those of the traditional methods. While both can produce credible results, the time, effort, and cost of the traditional methods could be put to better use for those factors that affect the results of the inclining more such as the inventory of weights to deduct.

Even if inclinings are required to use the traditional methods, a cheap electronic inclinometer should also be used as a back-up device so a full record of the entire evolution is available. Even a cell phone with an inclinometer application could be used for this purpose.

Today's technology simply provides a cheaper, more reliable, more accurate, more precise, and faster way to measure ship heel angle during an inclining experiment. It is time that certifying bodies accept this modern method.

References

[07916] *NAVSEA Technical Publication T9070-AF-DPC-010/079-1, Design Practices and Criteria for U.S. Navy Surface Ship Stability and Reserve Buoyancy*, January 2016.

[09616] *Naval Ship Technical Manual S9086-C6-STM-010, Chapter 096 Weights and Stability*, March 2016. Revision 3.

[Ame] American Society for Testing and Materials. *ASTM F1321-92, Standard Guide for Conducting a Stability Test (Lightship Survey and Inclining Experiment) to Determine the Light Ship Displacement and Centers of Gravity of a Vessel.*

[CT07] Dominick Cimino and David Tellet, editors. *Marine Vehicle Weight Engineering.* SAWE, Inc., May 2007.

[R D08] R Development Core Team. *R: A Language and Environment for Statistical Computing.* R Foundation for Statistical Computing, Vienna, Austria, 2008. ISBN 3-900051-07-0.

[Tel05] David Tellet. 3367. inclining experiment sensitivity analysis using excel simulation tools. In *64th Annual Conference, Annapolis, Maryland*, page 30, Annapolis, Maryland, May 2005. Society of Allied Weight Engineers, Inc., Society of Allied Weight Engineers, Inc. L. R. "Mike" Hackney Award.

[Tel14] David Tellet. 3617. in-service weight control for submarines. In *73rd Annual Conference, Long Beach, California*, page 23, Long Beach, California, 05/2014 2014. Society of Allied Weight Engineers, Inc., Society of Allied Weight Engineers, Inc.

[Tel17] David Tellet. 3678. Submarine Static Stability. In *76th Annual Conference, Montreal, Canada*, page 39, Montreal, Canada, 05/2017 2017. Society of Allied Weight Engineers, Inc., Society of Allied Weight Engineers, Inc.

About the Author: David Tellet has a Bachelor's degree in Ocean Engineering from Florida Institute of Technology (FIT) and a Master's in Systems Engineering from the University of Virginia. For thirty years he was a naval architect specializing in weights and stability and recently retired from the Naval Sea Systems Command as the technical warrant officer for submarine weights, stability, and equilibrium. He is a fellow and lifetime member of the Society of Allied Weight Engineers.

SAWE Paper No. S3772

Category No. 29

Scenario-based Prediction of Lightweight Costs - an Approach across Industries

Florian Wätzold, M. Sc.

Faculty V

Technical University Berlin

For Presentation at the

80th SAWE International Conference
on Mass Properties Engineering
October 2-6, 2021

Cocoa Beach, Florida

Table of Content

Abstract

The Paper:

To decide on which technology is best for a vehicle in development, technical and economical constraints need to be considered. The objective of this paper is to provide a generic, conceptual approach to estimating the value (€ or $) per weight unit (kg or lb), referred as lightweight cost, for all industries. It combines general project management and cost estimation techniques with mass property management. As cost and weight are unknown until the actual weighing or billing, this paper focuses on scenario-based assumptions. For an easy understanding, this lens is applied to a recently developed battery concept.

The described approach integrates basic project management processes such as risk management, estimation considerations and cost assessment. In this pursuit, mass and cost are rolled-up based on the breakdown structure. Taking the uncertainties into account a most likely, best, and worst case are evaluated and a cone of respective lightweight cost is generated. For the ease of use in industrial daily business exactly one value per pound for decision making is derived, condensing the lightweight cost range by superimposing the mass and cost according to their specific scenario probability.

The Author:

At the end of his study programme with Deutsche Bahn, Florian was working on concepts for the largest lost & found office in the world and implemented cost saving initiatives for logistics. While he grew a lot personally and professionally in these positions, he desired to learn more about engineering and decided to return to university to pursue a Bachelor and Master´s in mechanical engineering. Alongside studying, Florian started working with TGM Lightweight Solutions as a (jun.) consultant in 2018, mainly on projects to reduce the weight of future vehicle designs or implement mass property management processes with car OEM, but also for rail OEM and an Arianne 6 supplier. In spring 2021 he obtained his Master´s degree from Technical University Berlin.

"it is very difficult to predict, especially the future"

Niels Bohr

Introduction

Mass properties data are important not only for the effect of mass on the vehicle's performance and safety, but also for cost estimates, predictions, and cost management of the system or vehicle [1]. Within the aerospace industry, standards are established to manage mass properties of vehicles such as the ABSI/AIAA S-120A-2015 or SAWE Recommended Practice Number A-3. Such global standards do not exist for the automotive and rail industries. The objective of this paper is to develop a basis for a generic management system for all vehicles while linking weight and cost predictions as one approach of project management. This would be mainly part of the monitoring and controlling process group, whose responsibility it is to track, review and regulate a project´s performance or evaluate change requests [2].

Based on physical constraints, the weight criticality of any given (aeroplane) item depends on its weight and centre of gravity, the cost and time required to manufacture the item and its weight maturity [3]. Therefore, the design considerations need to take cost, manufacturing limitations, design feasibility and program schedule into account [3]. Mass properties management processes include estimating, calculating, measuring, managing, predicting, and reporting the vehicle's mass and other properties such as moments of inertia. These processes support the vehicle development from its inception throughout the vehicle's lifecycle, including disposal [1]. From an investment perspective, this results in costs and benefits both commonly considered as the value of a pound [1].

The exact lightweight cost depends on the location of the weight, the vehicle type, the design maturity and savings in manufacturing and operational issues [1].

In Figure 1 the situation is visualised. Starting from a known baseline weight a bandwidth of cost and weight is spanned depending on the tolerances and occurring opportunities or threats. Regarding the needed mass from a mechanical point of view, concepts like the weight funnel limit the weight range due to the convergency of consistent requirements and technology parameters [4]. Validating measurements are necessary due to insufficient modelling and imprecise measurements [5].

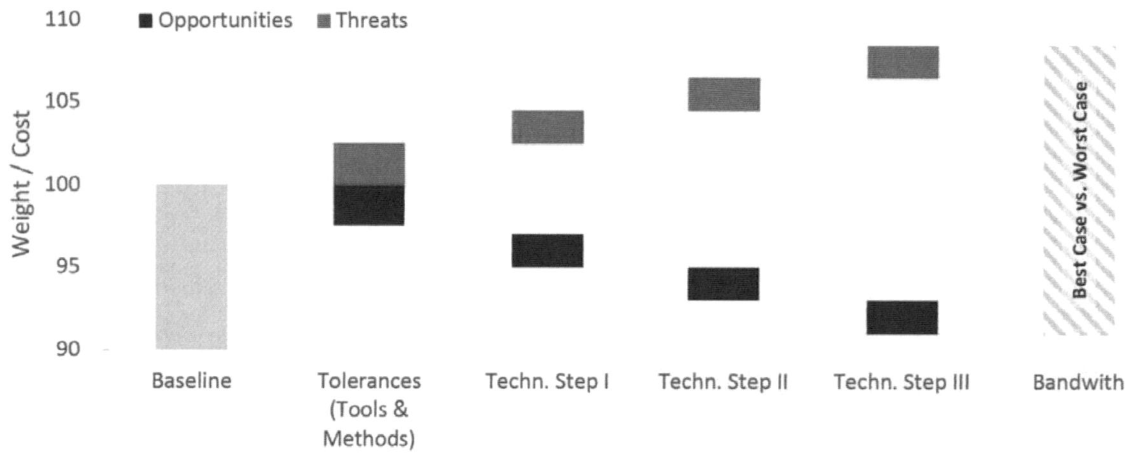

Figure 1: Bandwidth of weight and cost predictions [4]

SAWE recommends to complete the majority of weight optimization during early phases of design [3]. As the risk and the ability of all stakeholders to change product characteristics at low costs are highest at the start of the project [2], steering weight and cost issues as early as possible results in a reduction of the total cost. One major imitation to this ambition is the unpredictability of occurring risks. All the uncertain events and missing data concerning the physical condition of the parts and, in the end, of the vehicle itself are part of the risk register. Listed in this register are opportunities and threats [6].

Commonly, a work breakdown structure organises the work packages of a project and can also be used as a communication tool between engineering, management, and other functional teams [7]. An example of a work breakdown structure is given in Figure 2. A product-oriented work breakdown structure provides information on cost and technical performance on specific work packages. Whereas a functional work breakdown structure focuses less on these details. Hence, best practice for estimating cost is a product-oriented work breakdown structure based on specific hardware elements [7]. Weight and cost of an existing vehicle can then be derived by a bottom-up approach rolling up the weight and cost per item of the product column. But this fails, if no vehicle or parts are available because the vehicle is still in its early concept stage. Therefore, this paper sums up the fundamentals of project management and its interaction with basic mass property management (of aerospace applications) as well as cost estimation. All three topics are then merged into a generic approach to estimate the bandwidths of value of a pound prediction, referred as lightweight cost. The bandwidth depends on the uncertainties of the assumptions defined by different scenarios. Benefits of this analysis are its potential to increase performance of the vehicle or reduce its time to market and product cost.

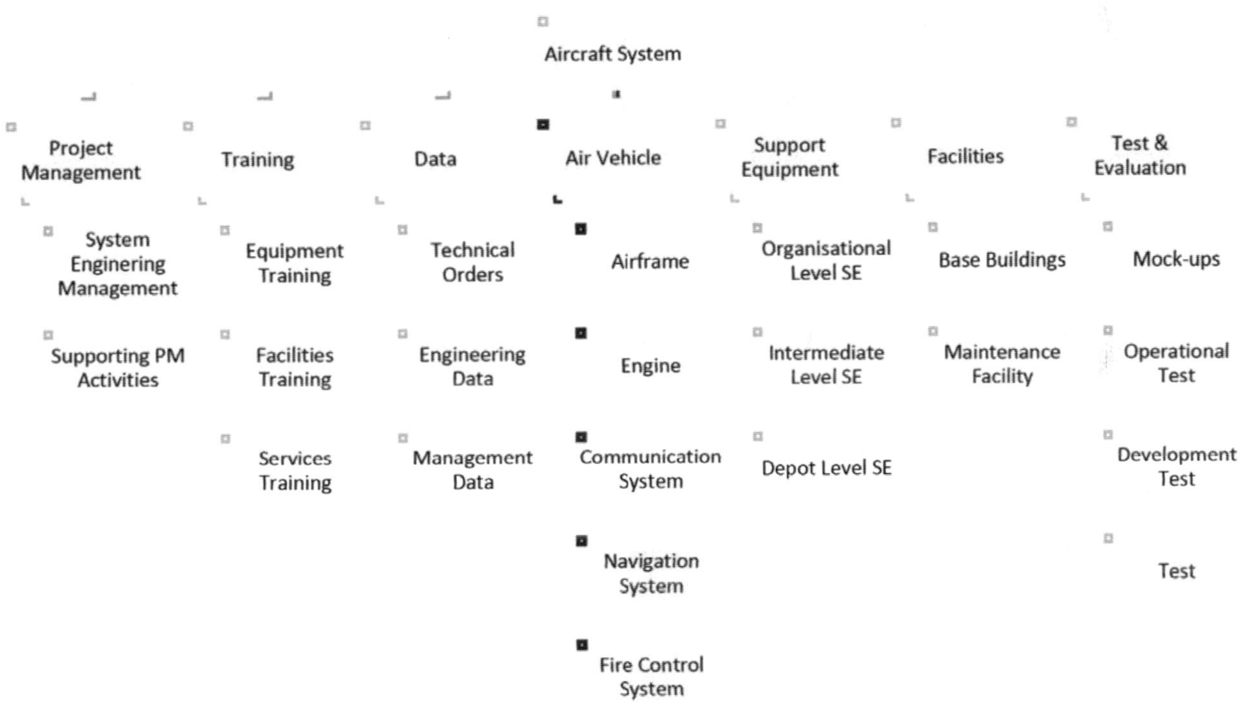

Figure 2: Example of a work breakdown structure [2]

Research

Project Management & Risk Management

Within the logical group of project management processes five groups can be distinguished. These are initiating, planning, executing, monitoring / controlling and closing [2]. As most projects strive for financial success, it becomes important to plan cost management, estimate the project costs, determine the budget and control the costs [2]. Especially the latter is linked to the planning of risk management, identification and analysis, response planning, response implementation as well as monitoring risks [2]. Similar to the mass margin allowance, a management reserve is the delta from project budget to cost baseline [2]. This reserve might be used to cover risk responses or other contingencies while captured opportunities resulting in cost savings might become additional profit [2].

To evaluate risks, numeric values for probability and impact might be assigned and multiplied to a score which can be evaluated with a scoring scheme [2]. Besides probability and impact other characteristics like controllability and urgency might be used to prioritise risks [8]. Usual responses to deal with threats are to escalate, avoid, transfer, mitigate or accept them [8]. Opportunities might be escalated, exploited, shared, enhanced, or accepted [8]. The necessary step before these undertakings is to assess the risks.

Estimating

According to PMI, a process for project estimates based on the PDCA circle is split in two iterative process groups. The first prepares the estimations and then manages and creates the estimates. The second creates estimates first and sets a baseline. Afterwards the actuals will be applied, reviewed, and controlled. This may lead to re-estimates [8]. The main methods of quantitative estimating approaches are analogous, parametric and bottom-up, [8] as well as relative estimating. However, usually several estimating approaches might be used to detect variations among them [8].

The first technique is a top-down approach based on similarity. Advantages are the minimum requirement of detailed information and focus on system level information. Down-sides might be inaccurate estimates and inappropriate historical examples [8]. The analogous approach uses ratio estimating [8], range estimating or three-point estimating [8]. The uncertainty of point estimates decreases over the project time as the general statements translate into detailed requirements, while reducing risks often also increases the potential cost [7]. In a more sophisticated form, three separate values (e.g. cost) are provided for the optimistic (A_o), pessimistic (A_p), and most likely (A_m) case. These three values are input to the beta distribution formula [8].

$$\beta = \frac{A_o + 4 \cdot A_m + A_p}{6}$$

Equation 1: Equation for beta-distribution [2]

A sufficient probability distribution might be a beta-distribution shown in Figure 3. The example contains 10 data sets each with a worst case of 240 units, most likely case of 120 units and a best case of 60 units. The standard derivation for each data set varies from 13 to 73 units. For this example, an 80 % confidence is assumed and therefore the lower-bound threshold is 1.118 and the upper-bound threshold is 1.482. Instead of calculating with 10 x 60 units (600 units), 120 (1,200 units), or 240 (2,400 units), the mean value according to Equation 1 is 130 units per data set or a total of 1.300 units. In other words, based on a specific scenario per data set the risk can be considered and summed up for various entries. Superimposing the values condenses the scenarios to one number of expected units in the blue area which are 80 % likely (1,118 to 1,482 units) and the red areas with a 10 % likelihood (below 1,118 units or greater than 1,482 units).

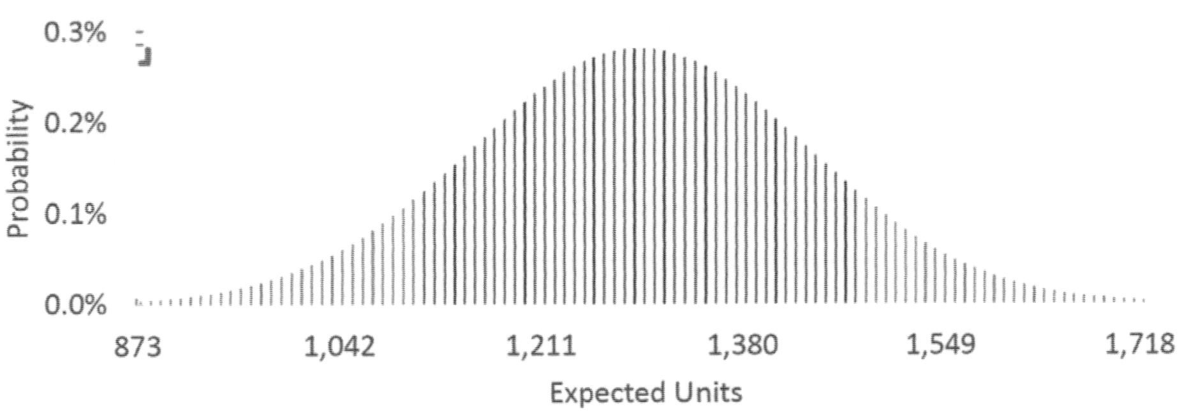

Figure 3: Normal Distribution (Statistical Program Evaluation and Review Technique)

The second technique to estimate is based on mathematical equations plus historical information of similar projects and taking scale differences into account. Estimating results are deterministic or probabilistic [8]. The benefits of it are higher accuracy and reliability compared to the analogous form at the cost of more estimating effort [8].

The third technique is deterministic and can be applied once project data becomes available. All available data will be rolled up to the higher levels of the work breakdown structure, resulting in the definitive estimate [8].

The third approach - relative estimating techniques - is required by the acceleration of change with no experience, data, or time available. These create high-level estimates, and it is recommended to use different units of measure such as T-shirt sizes [8]. Once enough data is available, the relative and traditional estimation can be combined while keeping the uncertainty in mind [8].

The creation of reliable assessments is challenging and driven by psychological reasonings such as optimism and pessimism [8] or anchoring to initially offered information [8]. These could be mitigated by the use of advanced analytics techniques based on artificial intelligence and machine learning [8].

Mass Property Management

A touch point of risk management and estimating is also mass properties management. Within the management processes, planning the technical effort of mass properties is followed by managing the mass properties risks, developing mass properties metrics and controlling the mass property baseline [1]. The responsibility of the engineer in charge is to understand the origin of the technical uncertainties, quantify them and prepare recommendations and decisions to handle those risks [1] during all project phases, so performance, schedule and cost are met [5]. The mass of a complete space vehicle might be broken down into a weight cascade of dry mass, predicted dry mass, predicted wet mass (at launch) and a ultimate wet mass at launch [5].

The approximate uncertainty of mass properties will be checked at specific project milestones and can be based on historical or comparative data or can be derived from physical assumptions. A key performance indicator, mass margin, shows the difference between allowable (expected) mass to predicted mass [9]. A mass growth allowance per category encompasses design uncertainties caused by the maturity levels [10]. Recommended determination categories for the maturity range from estimated (values based on preliminary data such as sketches) to calculated (values based on released drawings or CAD models) to actual (values based on measurement or comparison of similar components) to finally customer-furnished equipment or specification values [9]. The idea of quantifying weight with uncertainty levels is evident within all industries and project phases [11]. Figure 17 in the appendix contains an example from the standard for mass properties control for space systems for the mass growth allowance by design maturity.

All requirements are assessed and checked according to the thresholds for the mass margin allowance. The output of this process group is related to managing the threats and opportunities. All risks may be assigned a probability of occurrence. These are used as weighting factors to estimate the expected change and predict the forecast mass [9]. Examples of how to challenge the mass maturity are given in Figure 15 in the appendix. It should also be considered that additional weight savings may be achieved in load-bearing parts, when reducing the gross vehicle mass. A case-study of the MIT and the GM Research and Development Center locked four key subsystems (of 13 total), which resulted in a secondary mass saving potential to a mean of 0.12 kg/kg (lb/lb) with a 5th percentile of 0.1 kg/kg (lb/lb) [12].

One approach to quantify uncertainties and risk throughout the design development phase is selecting a work breakdown structure and assigning uncertainty categories to all mass properties entries. Within the next step, the average uncertainty based on historical data is quantified and a probability distribution per uncertainty category is used to set-up, for example, a Monte Carlo simulation. Finally the risk can be assessed by adjusting the mass properties variables [11].

Cost Estimating and Assessing

In addition to the technical performance of the product or concept, the financial aspects are important as well. The purpose of cost estimates is two-fold. The first is to evaluate the project's affordability as well as its performance against plans and the assessment of alternative systems or solutions. The second purpose is to provide the required budget and request for funding [7]. A specific activity of cost analysis is cost estimating [7] and combines, for example, concepts from accounting, engineering, mathematics, etc [7]. A thorough cost analysis takes engineering aspects (development, design, start up, initial vehicles, software, test and evaluation, special tooling, test equipment, facility changes), procurement and investments (site activation, training, equipment, initial, spare, repair parts, interim contractor support), operations and support through the entire life-cycle (manpower, fuel, maintenance, support, sustaining engineering) and disposal into account [7].

An important part of any engineering project is manufacturing costs. These are the sum of the cost of materials, tooling, fixed, capital and labour costs [13]. And are mainly driven by the production rate [13]. Usually, two estimating categories are used. Either the estimate is a life-cycle estimate including independent cost estimates and assessments as well as total ownership costs, or a business case analysis including an analysis of alternatives or economic issues [7]. An example of the life-cycle estimate is given in Figure 16 in the appendix. Besides the life-cycle cost, the business case analysis also focuses on other quantifiable and nonquantifiable benefits to ensure the decision is comprehensible for independent auditors reviewing it [7]. The output of estimating the cost can be used as input to select sources, assess technology changes, analyse alternatives, as well as perform design trade-offs [7].

To roll-up the cost estimates based on a product-oriented work breakdown structure, the lowest cost level is labour and materials which is supported by a detailed part list. To all these direct costs an overhead fee is added. From the project plan the quantity and schedule can be deducted to include the effects of learning and production rate [7]. Analogous to the weight maturity, the cost estimates normalising with the life cycle, effects from learning as well as production rate should be considered [7]. Regarding the latter Porter deducted: "The costs of a value activity are often subject to economies or diseconomies of scale. [...] [and] must be clearly distinguished from capacity utilization. [...] The scale sensitivity of activities varies widely. Value activities such as product development [...] are typically more scale-sensitive than activities such as procurement and sales force operations because their cost are heavily fixed no matter what the firm´s scale is" [14].

These learnings within the life cycle also result in a bandwidth of the cost estimate as shown in Figure 4. To assess the project´s budgeting, an uncertainty analysis with a cumulative probability distribution derived from a Monte Carlo simulation can be performed [7]. The results may be used to choose the level of contingency reserves needed to cover risks such as unexpected design complexity or incomplete requirements [7].

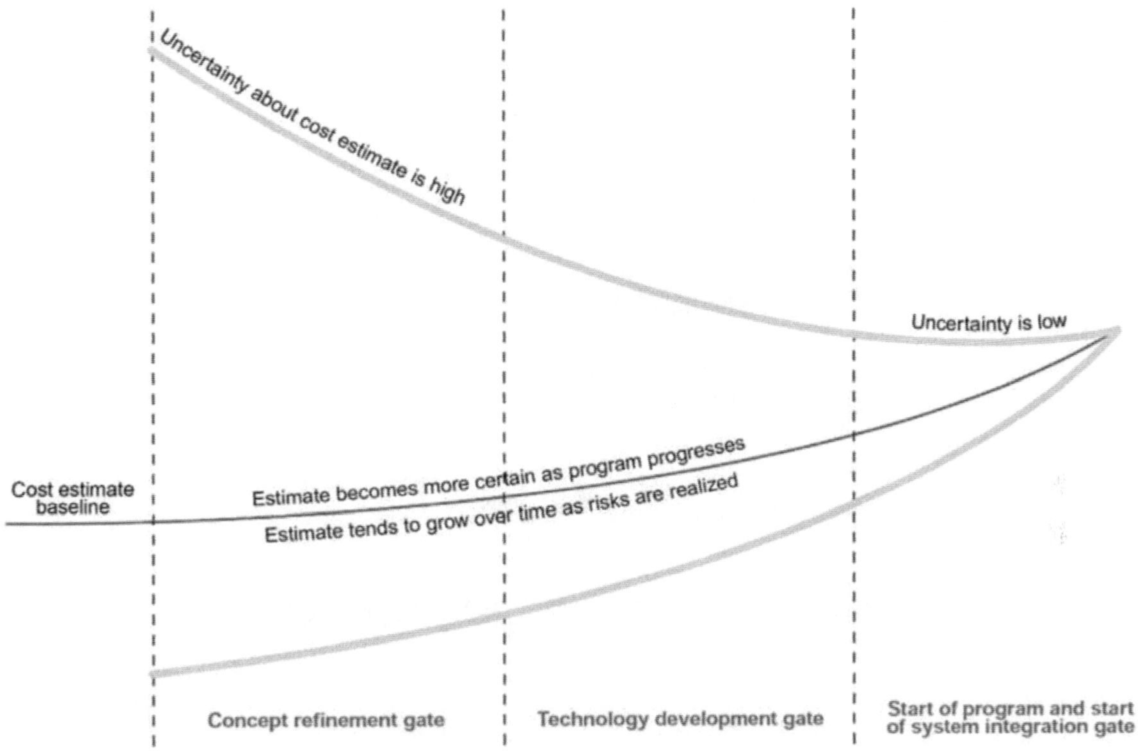

Figure 4: Cone of Uncertainty [7]

As the uncertainties per work breakdown structure element differ and may be asymmetrical, it is difficult to aggregate an appropriate probability distribution [7]. This can be solved by creating a statistical probability distribution with a specific risk shape per work breakdown structure element [7] and define its distribution model [7]. Examples for distribution models are given in in the appendix. Commonly used are the triangular, lognormal, beta, uniform, and normal distributions [7]. The normal or triangular distribution in combination with the mean and standard deviation are also a state of the art approach for evaluating mass properties risks [11]. In addition, it should be noted that correlations between different work breakdown structure elements´ costs may exist and create inconsistent scenarios leading to erroneous results [7].

Final outlook for this chapter is the overview provided in Figure 5. Due to economical constraints ranges for lightweight costs differ from 0 €/kg (0€/lb) for construction solutions to 3,000 €/kg (1361 €/lb) for applications in the space industry.

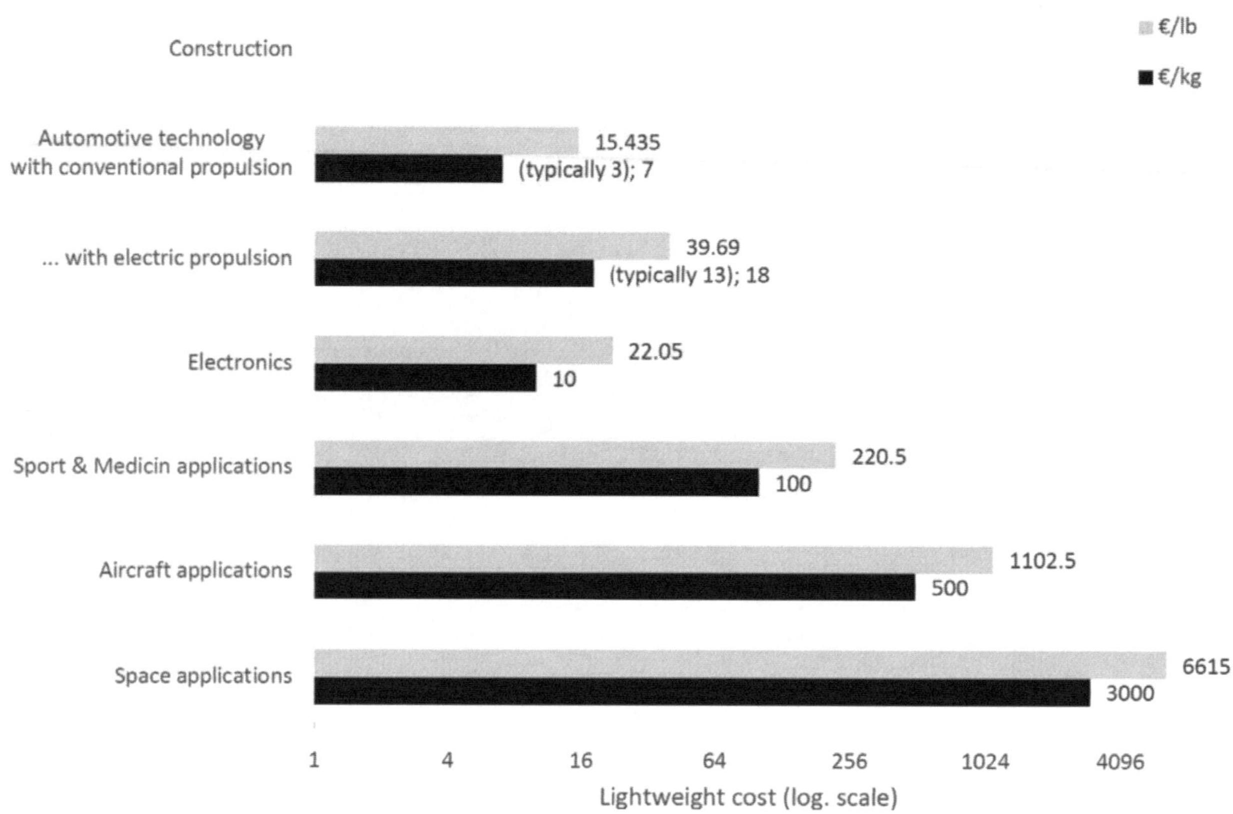

Figure 5: Ranges of lightweight cost [€/kg and €/lb] [15]

Discussion & Conclusion

Integrating Lightweight Cost into Project Management

A summary of the main activities is given in Figure 6 and might be integrated as part of the project management processes. When initiating the project, the basic assumptions are stated, and risk management is planned. To ensure the success of the project, lightweight costs should be focused on as early as possible. Based on the initial assumptions, the expected mass and cost can be derived, or an existing vehicle might be set as an estimating reference. If already needed, the centre of gravity and other mass properties can be set up as well. Within this framework the goals including the targeted lightweight cost may be set and synchronised with the project´s budget and mass properties restrictions.

In the elaboration phase, the weight estimates may be adjusted according to the details of the work breakdown structure and identified risks need to be analysed. At this point it should also be defined what use the component offers to the vehicle and which costs are relevant to the decision. For example, a technology change which substitutes two existing components by one may be benchmarked against the cost of both replaced components as well as related secondary weight and cost correlations. It is also important to clarify from the beginning which category of cost is important for the project. A vehicle sold to a mass market might be designed towards low direct cost during production to ensure a competitive price, while for other vehicle projects the total life-time cost might be relevant.

Uncertainties may be caused by insufficient mass modelling or imprecise mass measurements, currency exchanges, production rates and learning curves. It is obvious that some of these uncertainties will never be fully determined. In order to manage the lightweight cost the accountable project member or manager needs a clear understanding of the requirements and which non-conformance is critical to the vehicle and project. This becomes part of managing the risk during project phases. It should be noted here that neither the mass margin nor the financially driven management reserve is considered an actual margin. The core idea of both is to consider and manage the risks which will occur during the project. Therefore, the probability and impact of all risks are needed.

To assess the weight and cost predictions, a mass maturity analysis or cost analysis may be performed. Therefore, different approaches might be used to evaluate each component and its maturity level. A summary is given in Figure 7. Especially if the project deals with a completely new vehicle, a relative estimation might be the first step comparing single components to existing solutions.

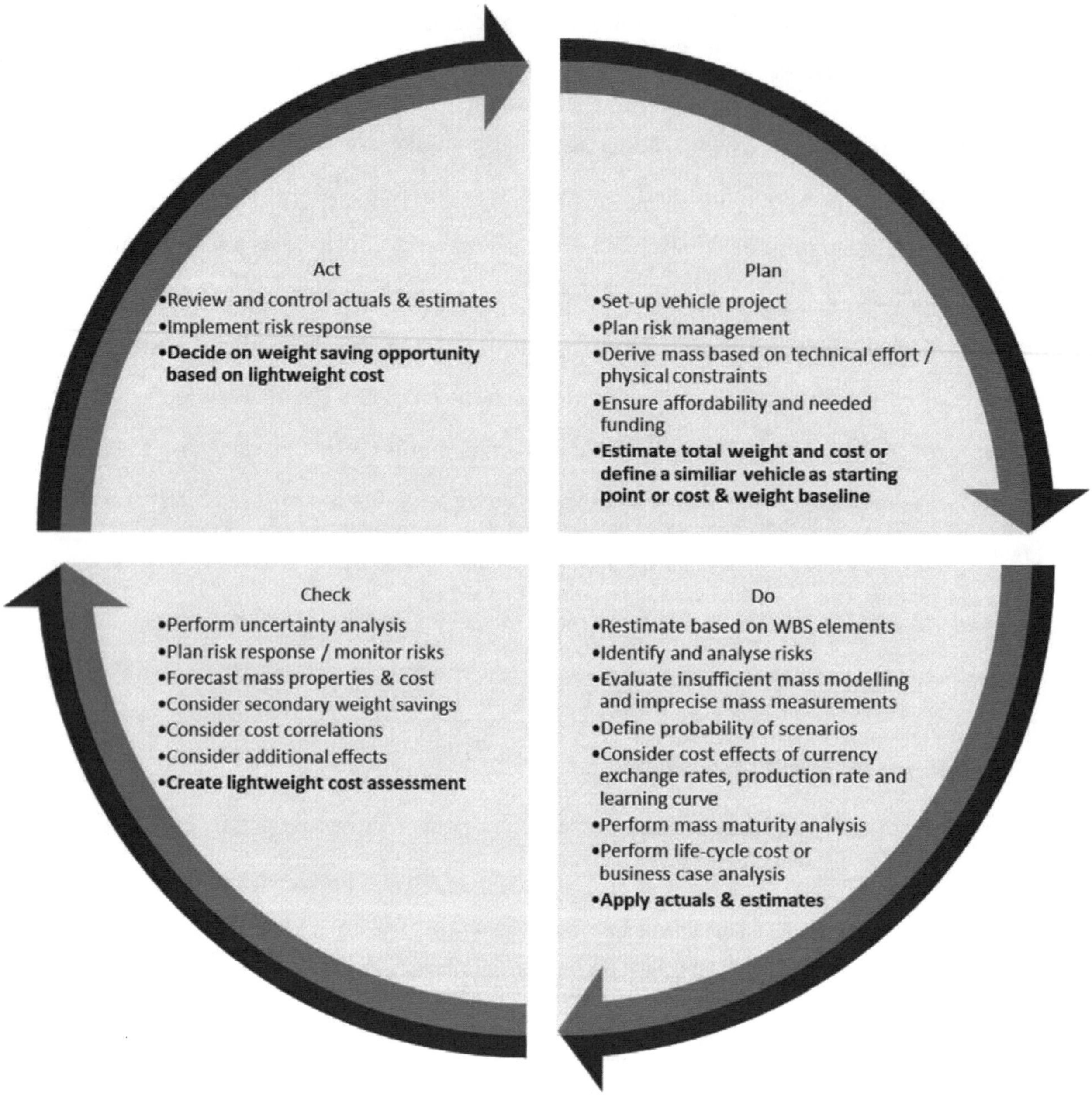

Figure 6: PDCA Cycle for Lightweight Cost

Once an understanding is created among the project members and relations to existing components are identified, weight and cost might be derived from these components. To add more details, physical and economical equations can be taken into account using historical data such as mass margin information. If a single deterministic value cannot be given, a probabilistic solution may be chosen. With a product-oriented breakdown structure in place, the estimations can by concretised based on the current layout, design or even by examining existing hardware solutions.

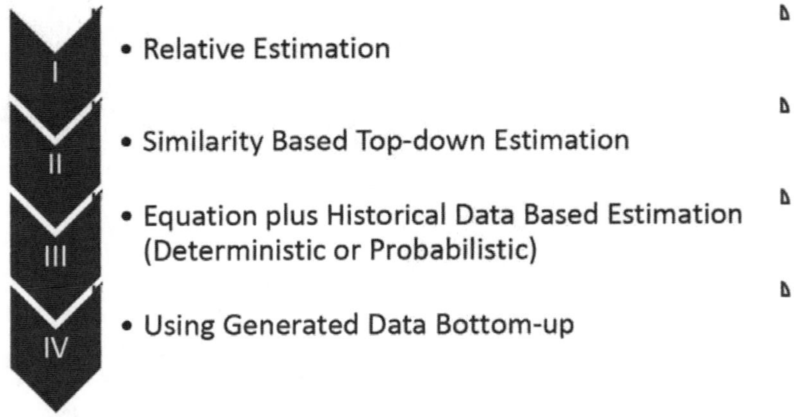

Figure 7: Estimation Approaches over Time

When performing the four steps of estimating, it is crucial for the involved members to focus on the probability of their assumptions. Output from the analyses are either actuals or estimates which update the expected cost and weight predictions. As the estimating process creates uncertainties due to unpredictable events and psychological effects, an uncertainty analysis should be performed. For this purpose, the work breakdown structure elements should be assessed individually to set the assumed probability distribution model and its boundaries for its mass and cost.

Within the project´s risk management, risk responses for the estimated values may be planned, and risks analysed and monitored. Based on all this known information, the mass property baseline and cost baseline may be forecasted. Secondary weight savings and cost correlations might be considered for the baseline values. All estimations and insights may be documented to be used in other projects. Finally, before restarting the PDCA cycle, the actual or estimated weight and cost are reviewed and controlled. This is also supported by implementing the planned risk response. Combining the information on weight and cost results in the lightweight cost assessment.

Lightweight Cost Assessment

To assess the lightweight cost the weight and cost of the current system needs to be measured against the proposed solution as shown in Figure 8.

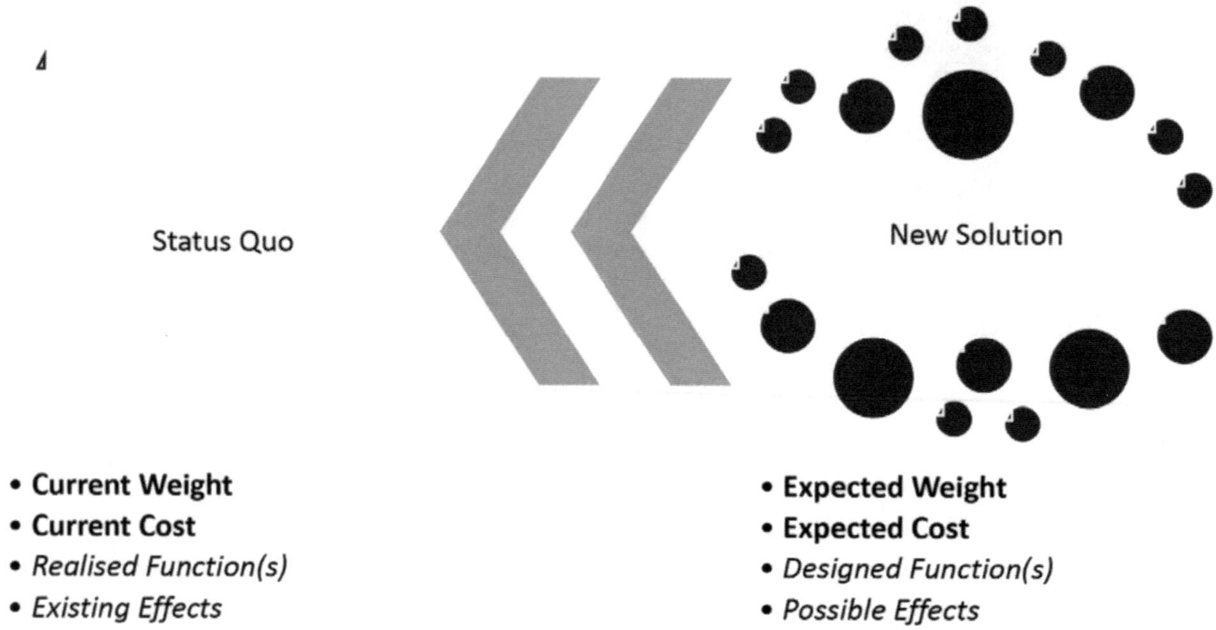

Figure 8: Evaluate Technology Change

In case of no available status quo, the defined baselines can be used for comparisons instead. Besides the direct cost and mass, other functions of the component or system might be taken into account to ensure secondary effects are considered as well. This might include all effects until elements are recycled and re-used in other applications. All upcoming correlations concerning the weight and mass should be noted or quantified for each idea. The basic calculation for the lightweight cost is to divide the current cost by the current weight and subtract the expected cost divided by the expected weight of the proposal.

As manufacturing costs define the core of the expected cost and the design, the evaluation might start by looking at the manufacturing method, production volume, material pre-selection, tooling, level of automatization as well as purchasing parts and needed licenses. Additionally, overhead fees, especially for engineering efforts, sales and logistic might be considered and a margin might be assumed.

Regarding the weight, all estimation activities might aim for creating a product breakdown and evaluate all parts as single items plus consider the integration into the final vehicle. For dimensioning, sketches and simplified mechanical substitute models and basic load-cases might be a sufficient starting point.

Cost and weight assumptions for defined production volumes might be categorized in best case, worst case and most likely case and an expected value can be created with a statistical approach using, for example, the introduced beta-distribution. Ideally a funnel with a shape comparable to the chosen distribution can be expected. All occurring discontinuities might be challenged.

An Example: A Modular, Automotive Battery Concept

Introduction to the Example

For understanding, what follows is an example that utilizes the described approach. The main assumption is the assessment of lightweight cost is part of the developing process for the hybrid BMW X5. As the main platform and cars design needs to be mostly identical to it, the shape and dimension of the vehicle can be borrowed from the X5 with internal combustion engine only. A draft of the hybrid car is given in Figure 9. The total length is 4886 mm (192 in), width of 2184 mm (86 in), height of 1762 mm (69 in) and a wheel base of 2933 mm (115 in).

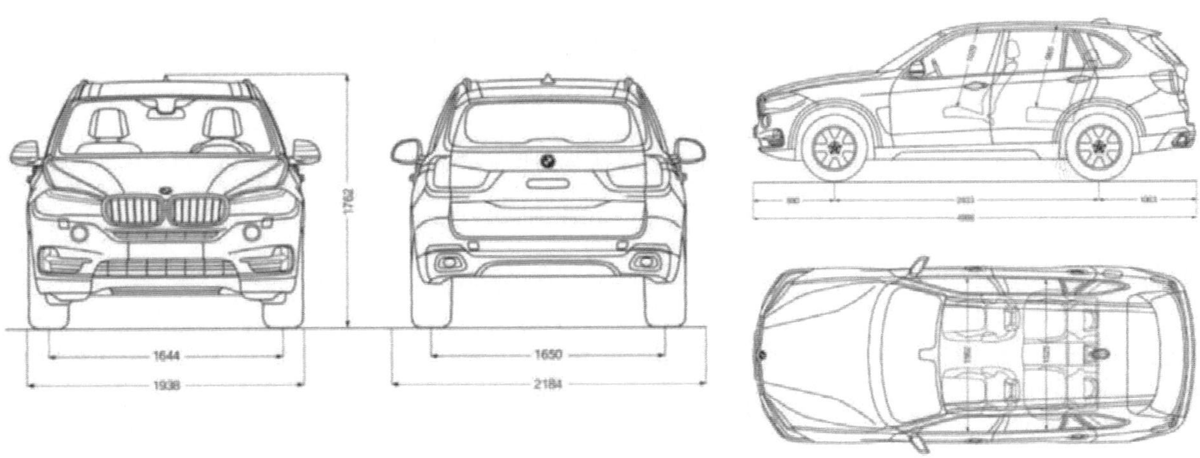

Figure 9: BMW X5 xDrive40e [mm] [16]

To give a best practice for a product related weight breakdown, an example for the physical X5 hybrid is given in Table 5 in the appendix. Additional information on the used LiFePO4 and manufacturing technology and the battery case concept (patent pending) is taken from a research project between TGM Lightweight Solutions GmbH (a lightweight consultancy), National

Manufacturing Scotland (research institute), Liteblox (battery manufacturer) and Koller group (an automotive tier1).

Estimation Loop I: Set-up & relative Estimation

It should be assumed BMW initiated the project and its governance properly and according to industry standards. Therefore, it can be deduced that, for example, risk management is planned and in place. To provide a baseline, a solution with two starter batteries (the actual engineered solution) comprised of a 90 Ah and a 50 Ah battery solution is defined. Because of this starting point a weight of 41 kg (90 lb) and cost of 80 € might be assumed. Due to other benefits, a threshold for the lightweight cost of a maximum of 18 €/kg (8.165 €/lb) as suggested by Klein will be assumed. Of course, there are multiple solutions possible to reduce the weight for starter batteries. For this example, a modular starter battery based on lithium iron phosphate will be set. Furthermore, only direct cost and weight effects will be taken into account. This increases the transparency of the suggested process.

This first step is supposed to be easy: to facilitate a creative mindset for analysing these lightweight costs within a brainstorming session. During such a session, first, ideas regarding risks should be noted. It is elementary to understand the maturity of the proposed solution, the ideas behind the ranking, controversies of the understanding and measurement, and any other effects of the technology swap. In Table 1, a relative estimation for the weight saving by substituting the lead batteries with lithium iron phosphate are provided. The ranking should be discussed by the group. In this case the better the rating, the more items are coloured in blue.

These kinds of batteries are already used in sports cars and therefore receive three out of three feathers. The downside is their high cost, though they might be cheaper as a supercapacitor. Thus, it is rated with one out of three piggy banks. The baseline solution is obviously not light (everyone can lift-up a regular starter battery which might be roughly 25 kg (55 lb)). However, it is a common and standardised solution and due to its economical advantages receives all three piggy banks.

Table 1: Relative estimation for the new battery concept

Component/System	Expected weight	Expected cost
Lithium iron phosphate battery		
Lead solution		

Estimation Loop II & III: Top-down Assessment & Equations Based Considerations

In this showcase, the boundaries between the top-down assessment and equation-based estimations are blurred. Consequently, both chapters are merged and as this paper deals with the approach, the technological aspects might be simplified.

Due to the fact that the gravimetric power of a high power lead battery (310 Wh/kg or 141 Wh/lb) is a quarter of the high power lithium solution (1,200 Wh/kg or 544 Wh/kg) [17], it can be assumed that the proposed battery solution might be roughly 3/4 lighter. The weight of the lithium battery might be a quarter of 41 kg (90 lb), or 10.25 kg (22.6 lb). Comparing the specific volumetric power, the ratio of a high-power lead battery (620 Wh/l or 2,347 Wh/gal) to a high power lithium solution (2,400 Wh/l or 9,085 Wh/gal) is fourfold. Therefore, regarding the weight of the battery case, it can be assumed that the new solution fits into the existing 90 Ah battery case, which has a mass of approximately 1.25 kg (2.8 lb) [17]. Summing up these thoughts results in an expected weight of 12.5 kg (27.6 lb).

Based on a assumption of 40 €/kWh for lead batteries solutions [18], a 7 % margin [19] and the casing´s weight being the main driver for the manufacturing cost, it can be deduced that the price for the 90 Ah case is 4.60 € [20]. For the cost of the lithium technology a cost range of 200 to 500 €/ kWh for $LiFePO_4$ [18] can be assumed. Two batteries with a total of 140 Ah powering a 12 V board net equals 1.68 kWh. Thus, a cost range of 336 € to 840 € for the cells can be expected. At this point it is not clear which cost can be expected as the range is 340.60 € to 844.6 €, including the 7% margin for the supplier. And obviously the situation for the weight is subject to discrepancies as well.

Without any historical data, the proposed beta distribution (Equation 1, page 8) is chosen from common models (see Table 4, page 28 in the appendix) as it can be used easily. Based on noted risks and critical thinking, the best, worst, and most likely cases might be defined in the group. A good starting point is the derived expected mass for the most likely case for the whole battery system. Without deeper understanding, the lower cost value might be assumed as the best case and the higher as worst case. A most likely case may be assumed in the middle or defined by an expert.

All gathered threats higher than a certain probable occurrence threshold should be added to come up with the worst cases. All positive chances higher than a certain probable occurrence threshold might be subtracted to generate the best cases. Maybe an extra adjustment factor or mass growth allowance is put on top of all scenarios to cover aspects that have not yet been considered. Also, all risks and weight snowball effects (e.g. a lighter battery holder because the battery is lighter), economical effects (e.g. learning curves), secondary cost savings (e.g. higher life expectancy) and additional values (e.g. less packaging needed) might be noted.

In Table 2 some values for the weight and cost of each scenario are summed up. For this example, let assume, Table 2 is the result the team agreed upon, after reviewing all the technical details (e.g. the weight drops even more as lithium iron phosphate provides a stable voltage over capacity increasing the effective capacity, one more component will be need to control the battery, ...). Please note that a total cost of 80 € for the existing batteries was assumed to calculate the lightweight cost.

Table 2: Loops II&III - Assuming scenario-based weights and costs for the new battery concept

	Best scenario		Most likely scenario		Worst scenario	
Values	5 kg (11 lb)	340.60 €	10 kg (22 lb)	592.60 €	15 kg (33 lb)	844.6 €
Lightweight cost	7.24 €/kg (3.28 €/lb)		16.54 €/kg (7.50 €/lb)		29.41 €/kg (13.34 €/lb)	

It is important to consider the weight and cost of the status quo when calculating the lightweight cost. According to the beta distribution chosen for the estimation process, all three light weight costs are superimposed to an expected value of 22.64 €/kg (10.27 €/lb) in estimation loop III. Although this is above the set threshold, it is assumed that the project manager asks for more details and starts the estimation loop IV to understand the drivers. It is possible that the most likely scenario which is in the range seems probable or because of scouting future trends for the vehicle generation.

Estimation Loop IV: Concept Model & Bottom-up Assessment

Based on exchange with cell suppliers or engineers in charge of designing batteries, the number of specific cells and the necessary battery management system can be determined. On that basis a CAD model might be derived and the components´ weight be calculated. As the design matures, the discussions and requirements become more sophisticated. An example for estimating the scenarios is given in Figure 10 (Figure 22 for lb).

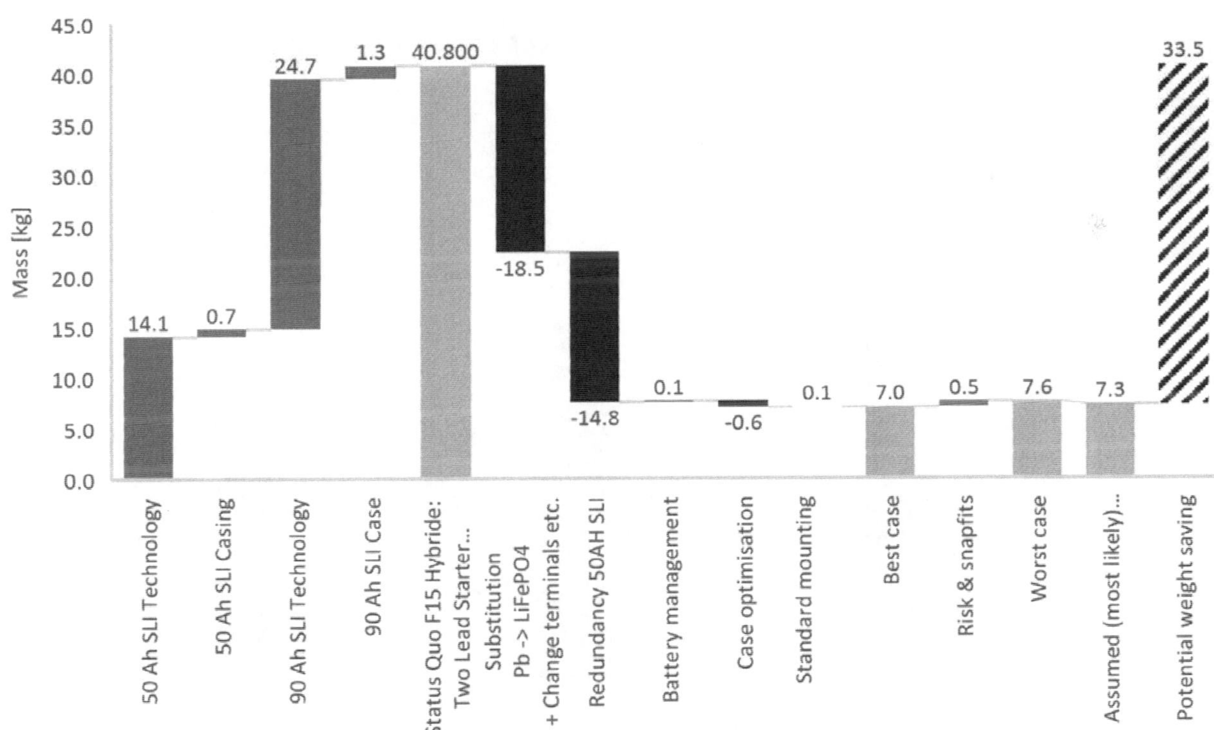

Figure 10: Waterfall chart of the weight saving potential [kg] [21]

It can be seen that the weight savings is higher than the rough calculation in loop II. Estimating the weight by the gravimetric power supply neglects, for example, the redundancy of one complete case.

With purchasing parts identified and a manufacturing scenario determined, the most interesting driver of the cost is the production volume. It is also important to keep in mind that the engineering process, tests, and other costs should be recouped from the product as well (see also Figure 16). Therefore, a cost model for the concept design and its manufacturing processes need to be generated. An example for a thorough cost analysis of the battery concept is taken from a paper about estimating the cost for this concept published in the Association of German Engineers' paper VDI-Z [21]. In this scenario standard injection moulding is set. Experienced engineers and a software solution, SEER for Manufacturing by Galorath, are used to consider all the primary time and cost drivers. These includes tooling, material pricing and capacity of the machines as well as assembly and inspection. During this phase, the concept might be optimised in iterative steps based on the derived insights. The outcome of this process for different production volumes is shown in Figure 11. Other outcomes of this thorough and holistic analysis might be noted or integrated into the risk management and sharpen the understanding of the assumptions and scenarios of this project.

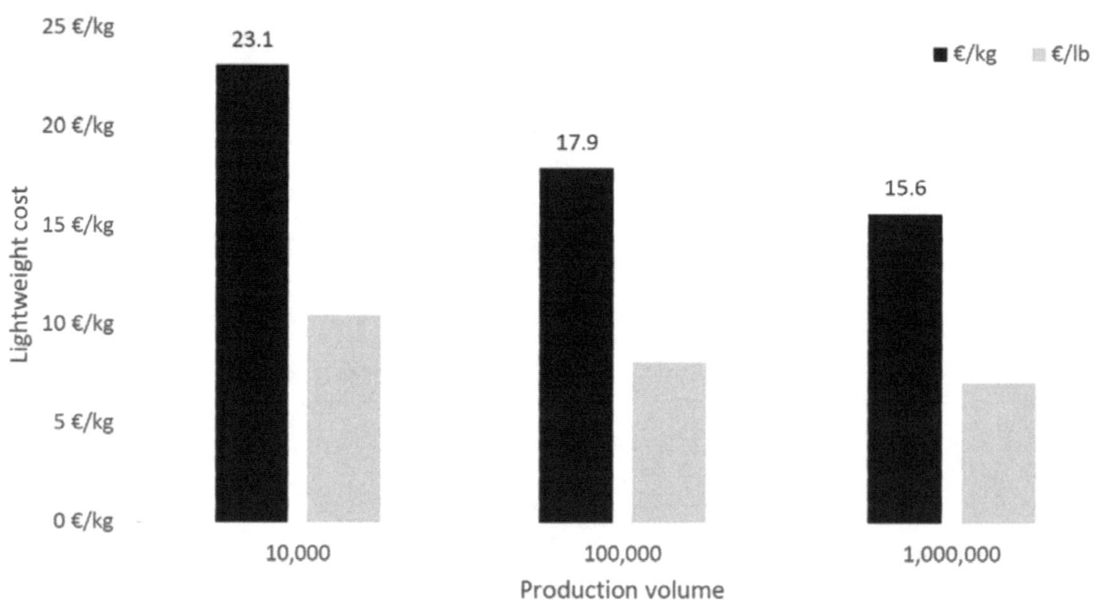

Figure 11: Lightweight cost over production volume [21]

The final numbers, determined by costs in Figure 12, are shown in Table 3. In comparison to the prior loop, the estimated weight range is decreased from 10 kg to 0.6 kg (22.05 lb to 1.32 lb), narrowing the expected mass to close to 7 kg (15.43 lb). Regarding the cost, the range is also decreased. To simplify this example, the cost difference for the scenarios depends on the production volume only.

Table 3: Loop IV - Assuming scenario-based weights and costs for the new battery concept

	Best scenario		Most likely scenario		Worst scenario	
Values	7 kg (15.43 lb).	610.71 €	7.3 kg (16.09 lb).	688.54 €	7.6 kg (16.76 lb).	864.74 €
Lightweight cost	15.63 €/kg (7.09 €/lb)		17.92 €/kg (8.13 €/lb)		23.11 €/kg (10.48 €lb)	

Conclusion: The development of the Lightweight Cost during the Design Process

Within every loop all estimations might be checked for insufficient mass modelling, imprecise mass measurements, and the need for other mass properties or the application of a mass growth allowance. Additionally, it is necessary to update assumptions and determine the maturity and uncertainty of the calculated results and designs. Changes might be handled when steering risks and implementing specific responses to those risks.

Figure 12, Figure 13 and Figure 14 show the result from the case-study from loop I to loop IV. The beta-distribution for weight and cost and therefore also for the lightweight cost is assumed according to Equation 1. The most important takeaway from these charts is to understand how the variance of the weight values declines as the concept matures.

For the weight the expected values stays within the weight funnel and starting from three feathers increases over 31 kg to 33,7 kg (68.34 lb to 74.30 lb) expected weight saving as shown in Figure 12 (Figure 23 for lb).

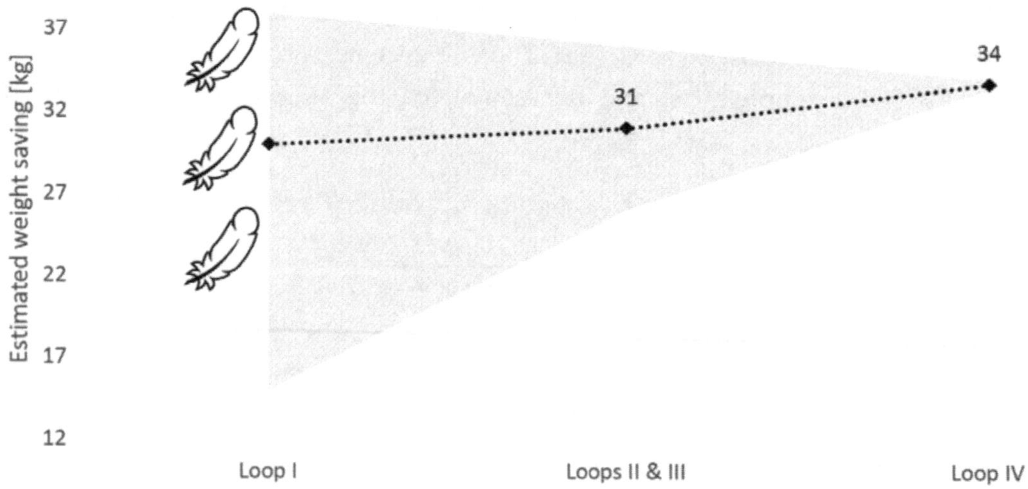

Figure 12: Weight funnel of the new battery concept [kg]

Figure 13 shows the cost estimation dipping from one piggy bank to 532 € per battery and then rises with the maturity to 705 €.

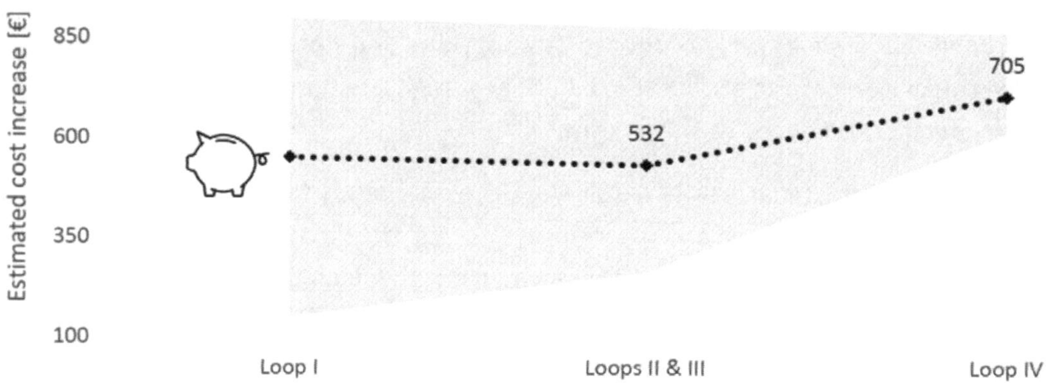

Figure 13: Cost funnel of the new battery concept [€]

At last the important indicator of the lightweight cost is derived on the estimated weight and cost. The result is shown in Figure 14 (Figure 24 for €/lb). Despite increasing cost, the more increasing weight estimation causes a decreasing lightweight cost. Starting from one of three piggy banks for three of three feathers the lightweight cost estimation converges towards the limit of 18 €/kg (8.17 €/lb). Based on the insights the decision can be made by the project leader.

Finally, if the lightweight saving is applied to a vehicle, the last step (e.g. loop V) is weighing the actual part. Any difference to all prior steps might be noted and considered for the project´s lessons learnt. The same value can be also realised when the project is closed and the vehicle´s lifetime is over to narrow down the cost spent on this lightweight saving. It is important to understand that even these numbers might not have one precise weight and cost. Measurement and manufacturing tolerances as well as rough allocation of cost might cause inaccuracies.

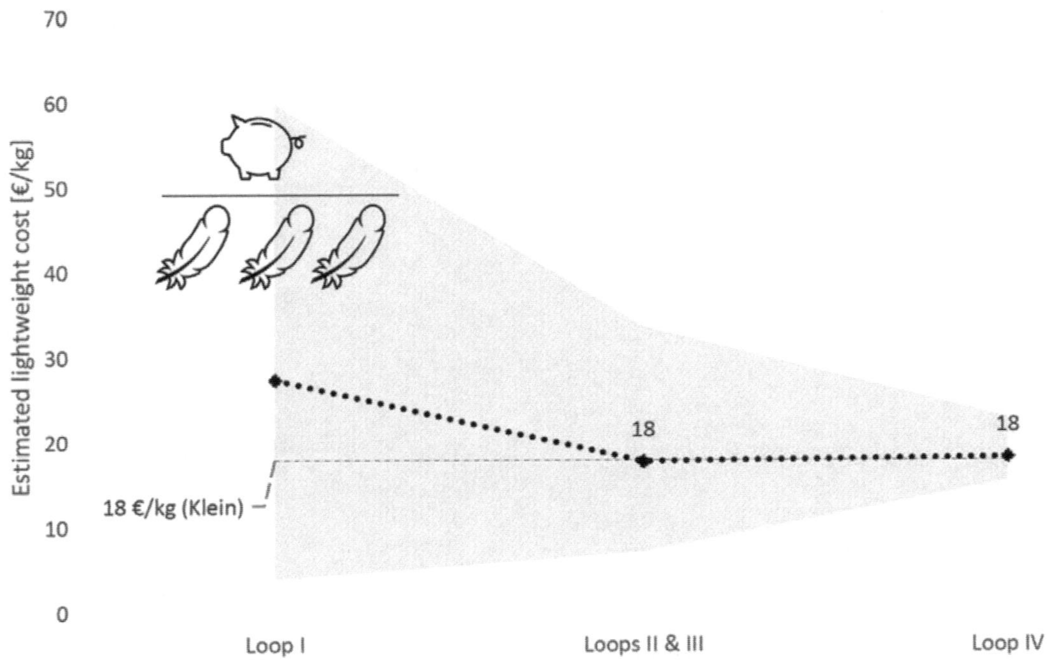

Figure 14: Lightweight cost funnel for new battery concept [€/kg]

The integration of the narrowing-down process as part of the PDCA cycle feeding the risk management with the underlying assumptions enables creativity and transparency while challenging the status quo. By doing so, the risk of anchoring to any initially provided value can be mitigated. This iterative approach and the simplified example provides a process to combine weight and cost estimations, specifying them and superimposing the different assumptions (building up the scenarios) to one comprehensive value. Thus, cost and weight can be analysed over their full range while still considering other possibilities. As the core process is not limited to a specific model to estimate weight, cost or stochastic issues, the project team might complement the idea by using more sophisticated methodologies to deepen the understanding of the weight saving opportunity and its probability of occurrence. An interesting field to explore is understanding the correlations of cost and weight. A valuable addition might be a reliable consideration of secondary effects for the vehicle´s weight and cost when changing any component or system. An example from the case-study might be assess the less needed package as well.

One benefit of this described approach is to follow up on any idea which might reduce the weight even if only rough details are know. A condition like this might be found at the beginning of any vehicle project and a structured process to evaluate lightweight cost is helpful to decide on specific weight saving action as well as trace already determined ideas when project constraints or assumptions changed.

Appendix A: Theoretical Background

Component´s heritage

- Is it a new modified or existing design?
- How extensive are the changes compared to existing items?
- Has it been qualified?

Status of system's design and analyses.

- Is it a design or estimate?
- Are load cases analysed?
- Are requirements such as mass, strength, natural frequency, thermal, shock balance, performance and environmental requirements met?
- Is it a layout or ready for manufacturing?
- Has it been built and verified?

Maturity level

- Estimated
- Calculated by parametric prediction or hand calculation
- Analysed based on Computer-aided-design (CAD) analyses
- Actual weight
  - Of a similar unit.
  - Of an identical unit.
  - Of the exact unit.

Figure 15: Mass Maturity Assessment [5]

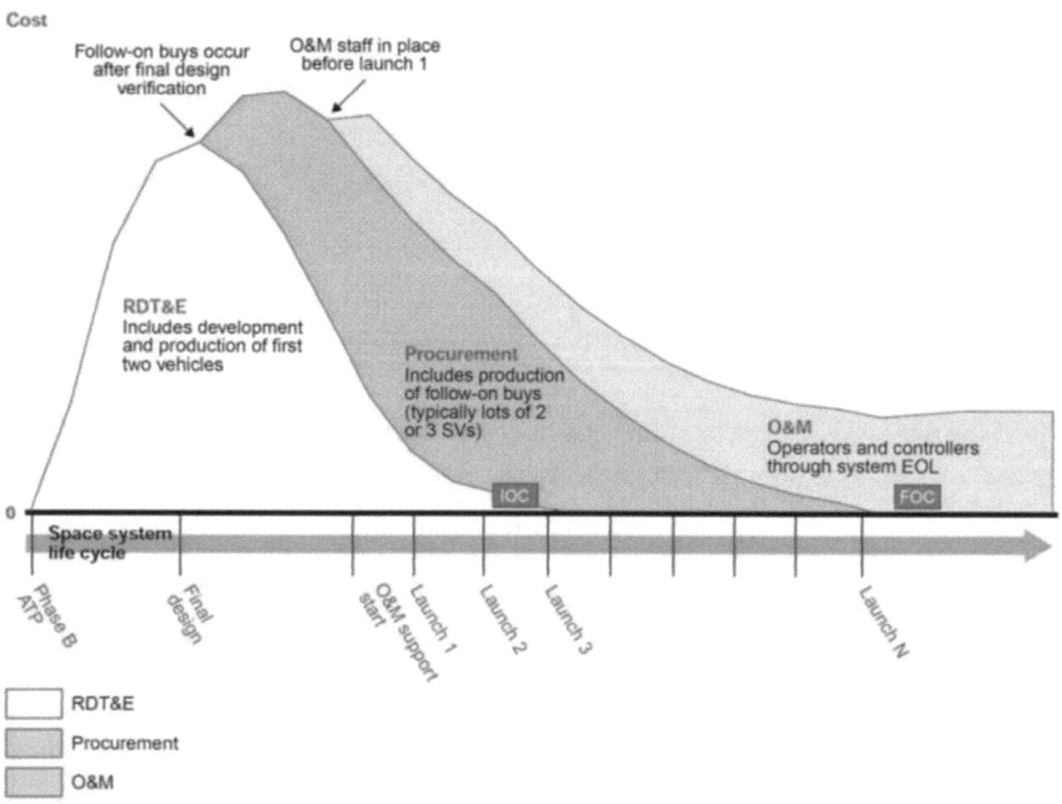

Source: DOD.

Note: O&M = operations and maintenance; RDT&E = research, development, test, and evaluation; SV = space vehicle; EOL = end of life; IOC = initial operational capacity; FOC = full operational capacity.

Figure 16: Life-cycle cost estimate for a space system [7]

Maturity Code	Design Maturity (Basis for Mass Determination)	Percentage Mass Growth Allowance												
		Electrical/Electronical Components			Structure	Thermal Control	Propulsion	Batteries	Wire Harnesses	Solar Array	ECLSS, Crew Systems	Brackets, Clips, and Hardware	Mechanisms	Instrumentation
		0-5 kg	5-15 kg	>15 kg										
E 1	**Estimated**	20-35	15-25	10-20	18-25	30-50	15-25	20-25	50-100	20-35	20-30	20-35	18-25	25-75
2	**Layout**	15-30	10-20	5-15	10-20	15-30	10-20	10-20	15-45	10-20	10-20	10-25	10-20	20-30
C 3	**Preliminary Design**	8-20	3-15	3-12	4-15	8-15	5-15	5-15	10-25	5-15	5-15	8-15	5-15	10-25
4	**Released Design**	5-10	2-10	2-10	2-6	3-8	2-7	3-7	3-10	3-5	3-8	3-8	3-4	3-5
A 5	**Existing Hardware**	1-5	1-3	1-3	1-3	1-3	1-3	1-3	1-5	1-3	1-4	1-5	1-3	1-3
6	**Actual Mass**	Measured mass of specific flight hardware; no MAG; use appropriate measurement uncertainty.												
S 7	**CFE or Specification Value**	Typically, an NTE value is provided, and no MAG is applied												

Expanded Definitions of Maturity Categories		
E1	**Estimated**	a. An approximation based on rough sketches, parametric analysis, or incomplete requirements b. A guess based on experience c. A value with unknown basis or pedigree
E2	**Layout**	a. A calculation or approximation based on conceptual designs before initial sizing (equivalent to layout drawings) b. Major modifications to existing hardware
C3	**Preliminary Design**	a. Calculations based on new design after initial sizing but before final structural, thermal, or manufacturing analysis b. Minor modification of existing hardware
C4	**Released Design**	a. Calculations based on a design after final signoff and release for procurement or production b. Minor modification of existing hardware
A5	**Existing Hardware**	a. Measured mass from another program, assuming that hardware will satisfy the requirements of the current program with no changes b. Values substituted based on empirical production variation of same or similar hardware or measured mass of qualification hardware c. Catalog values

Figure 17: Mass Growth Allowance by Design [9]

Table 4: Eight Common Probability Distributions [7]

<table>
<tr><th></th><th>Description</th><th>Shape</th><th>Typical application</th></tr>
<tr><td>Bernoulli</td><td>Assigns probabilities of “p” for success and “1 – p” for failure; mean = “p”; variance = “1 – p”</td><td>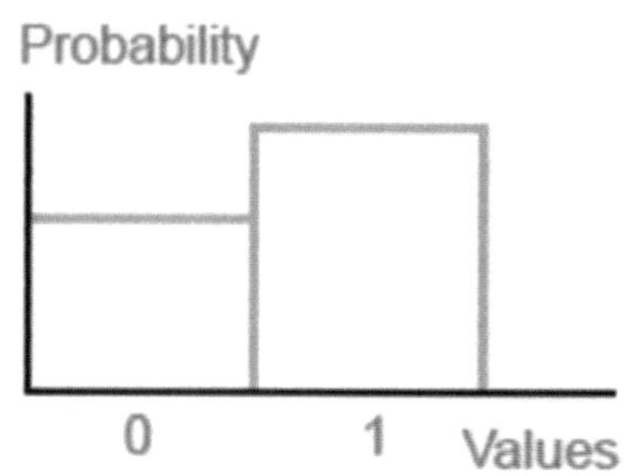
</td><td>With likelihood and consequence risk cube models; good for representing the probability of a risk occurring but not for the impact on the program</td></tr>
<tr><td>Beta</td><td>Similar to normal distribution but does not allow for negative cost or duration, this continuous distribution can be symmetric or skewed</td><td>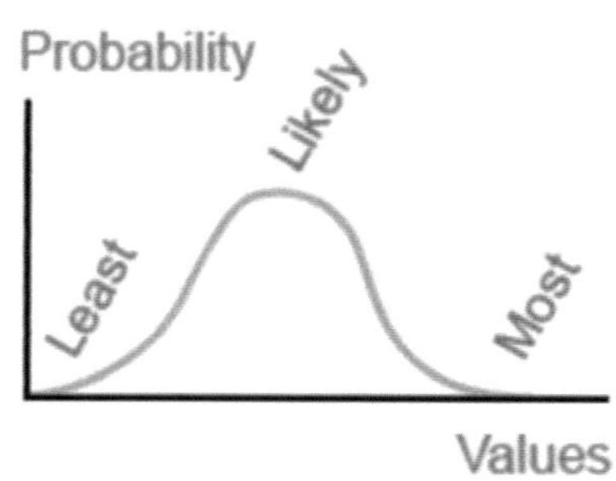
</td><td>To capture outcomes biased toward the tail ends of a range; often used with engineering data or analogy estimates; the shape parameters usually cannot be collected from interviewees</td></tr>
<tr><td>Lognormal</td><td>A continuous distribution positively skewed with a limitless upper bound and known lower bound; skewed to the right to reflect the tendency toward higher cost</td><td>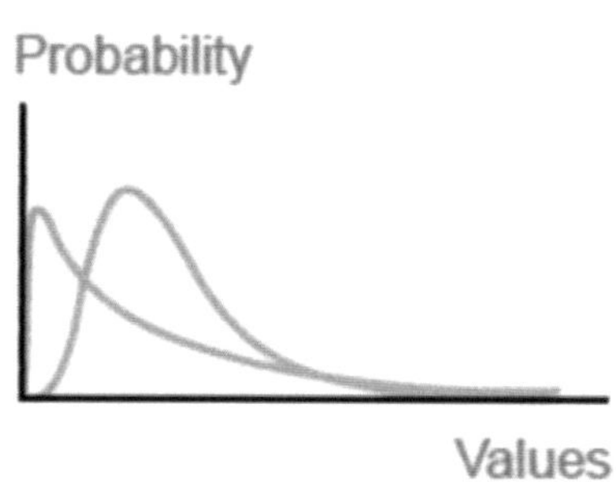
</td><td>To characterize uncertainty in nonlinear cost estimating relationships; it is important to know how to scale the standard deviation, which is needed for this distribution</td></tr>
<tr><td>Normal</td><td>Used for outcomes likely to occur on either side of the average value; symmetric and continuous, allowing for negative costs and durations.
In a normal distribution, about 68% of the values fall within one standard deviation of the mean</td><td>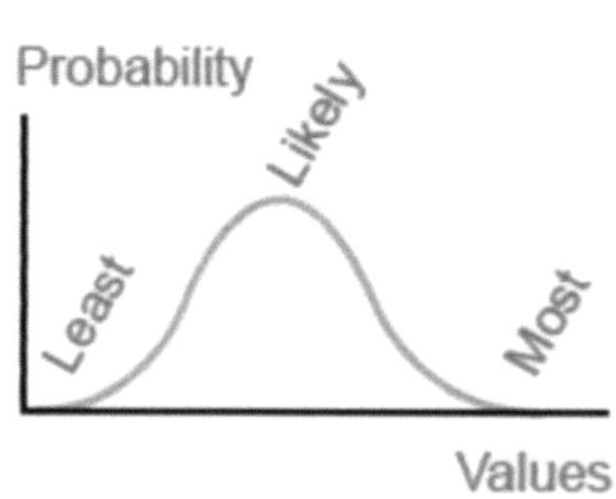
</td><td>To assess uncertainty with cost estimating methods; standard deviation or standard error of the estimate is used to determine dispersion. Since data must be symmetrical, it is not as useful for defining risk, which is usually asymmetrical, but can be useful for scaling estimating error</td></tr>
</table>

	Description	Shape	Typical application
Poisson	Peaks early and has a long tail compared to other distributions		To predict all kinds of outcomes, like the number of software defects or test failures
Triangular	Characterized by three points (most likely, pessimistic, and optimistic values), can be skewed or symmetric and is easy to understand because it is intuitive; one drawback is the absoluteness of the end points, although this is not a limitation in practice since it is used in a simulation		To express technical uncertainty, because it works for any system architecture or design; also used to determine schedule uncertainty
Uniform	Has no peaks because all values, including highest and lowest possible values, are equally likely	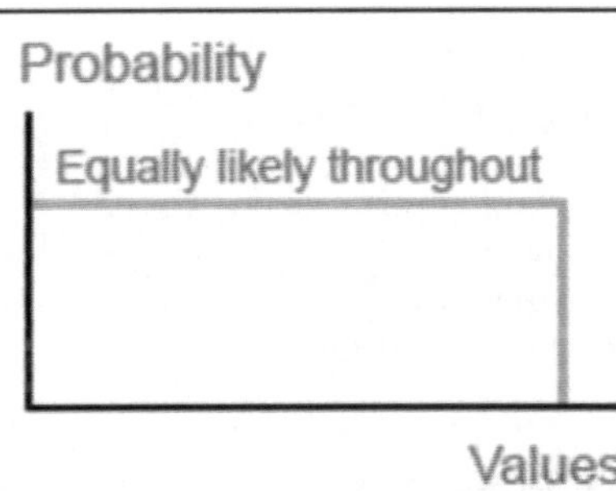	With engineering data or analogy estimates
Weibull	Versatile, can take on the characteristics of other distributions, based on the value of the shape parameter "b"— e.g., Rayleigh and exponential distributions can be derived from it (note: The Rayleigh and exponential distributions are a class of continuous probability distribution.	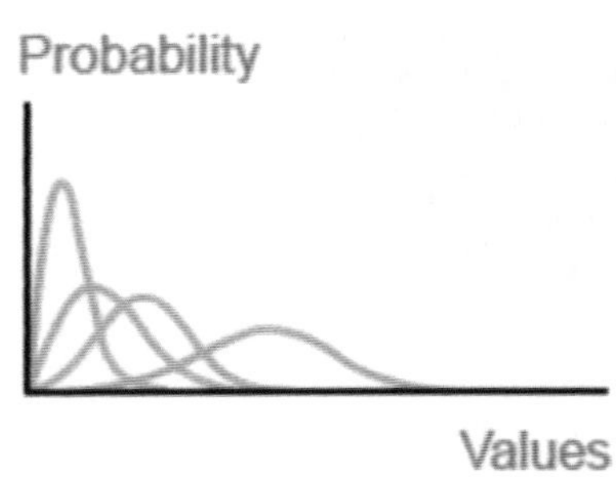	In life data and reliability analysis because it can mimic other distributions and its objective relationship to reliability modeling

Appendix B: Case-study Data

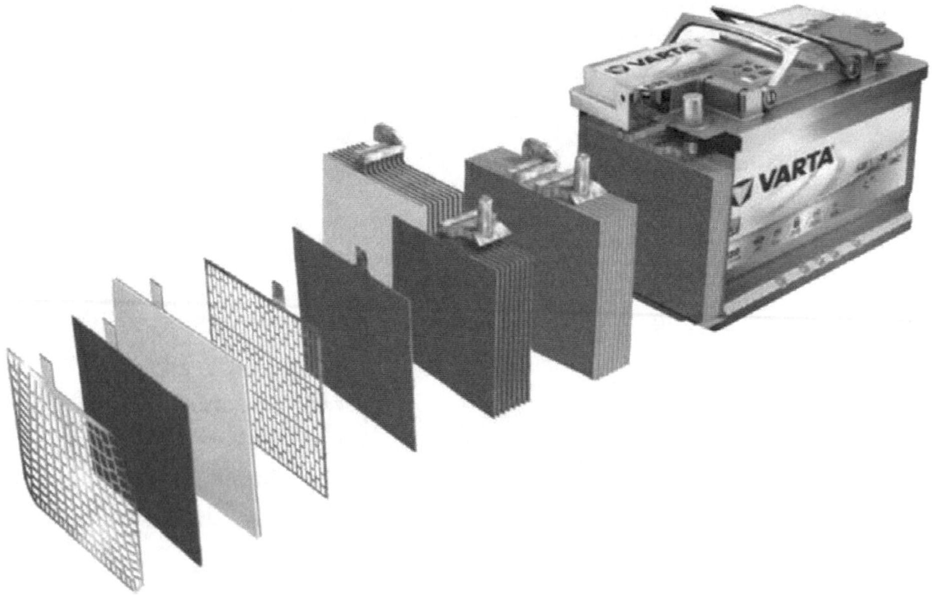

Figure 18: Rendered Cut Section of a State of the Art Starter Lightning Ignition Battery [22]

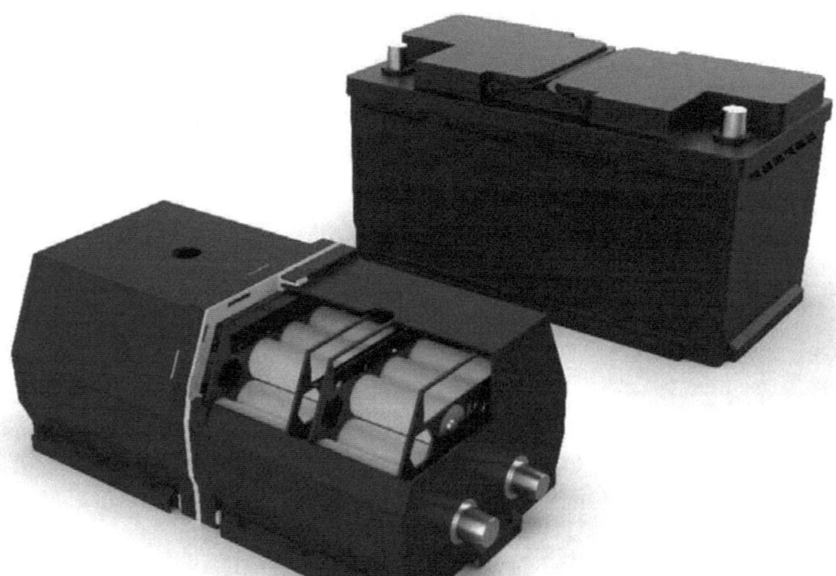

Figure 19: New Lightweight Battery Concept [21]

Figure 20: Powertrain BMW X5 40e [23]

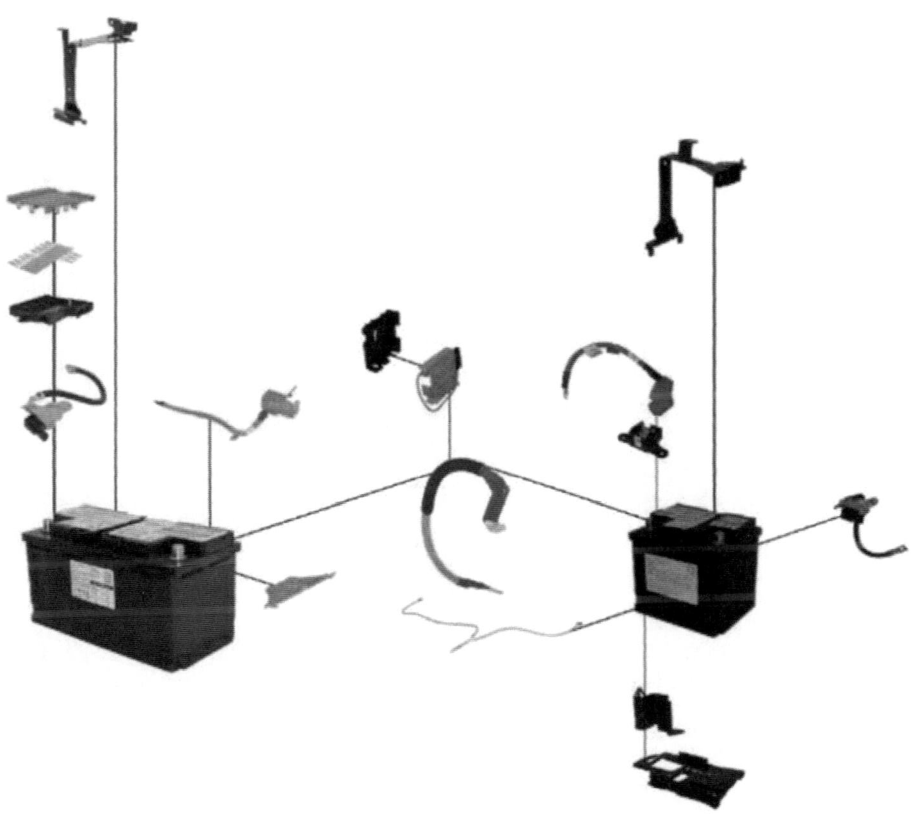

Figure 21: Complete Starter Battery System [24]

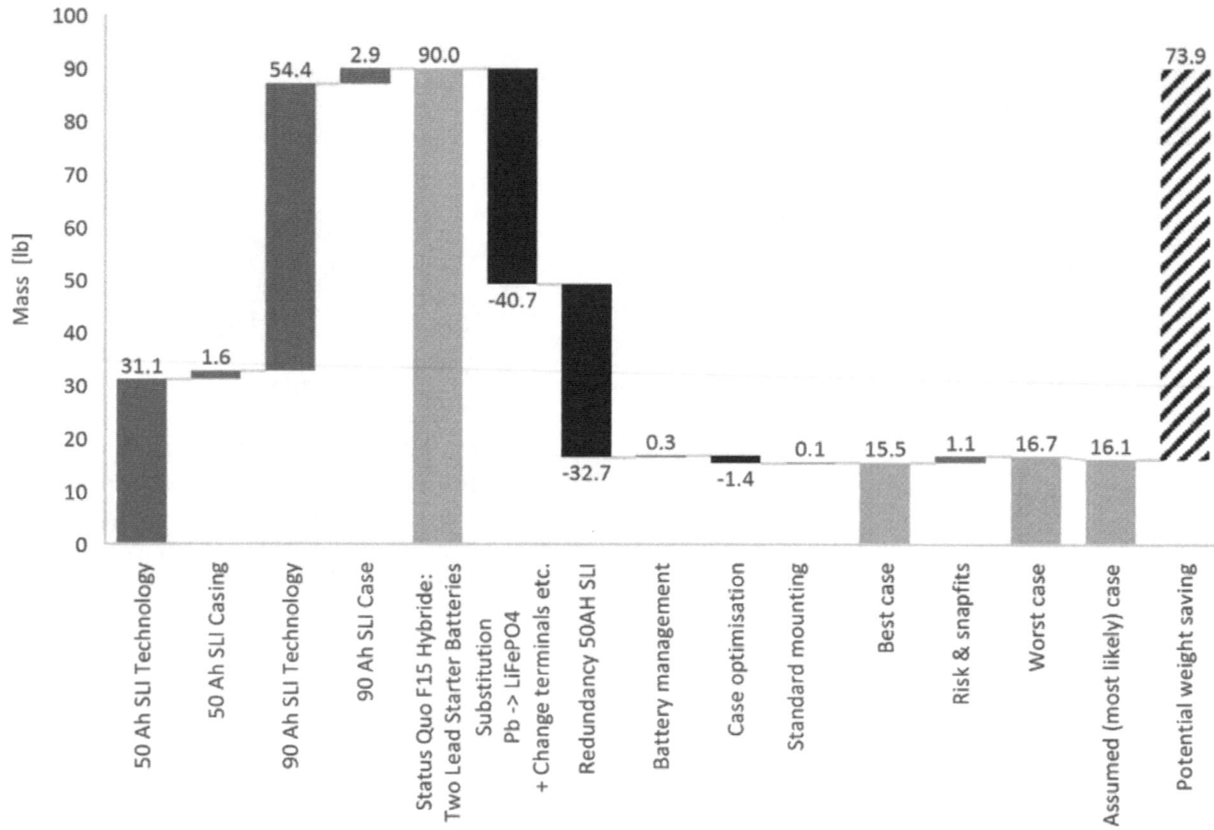

Figure 22: Waterfall chart of the weight saving potential [lb] [21]

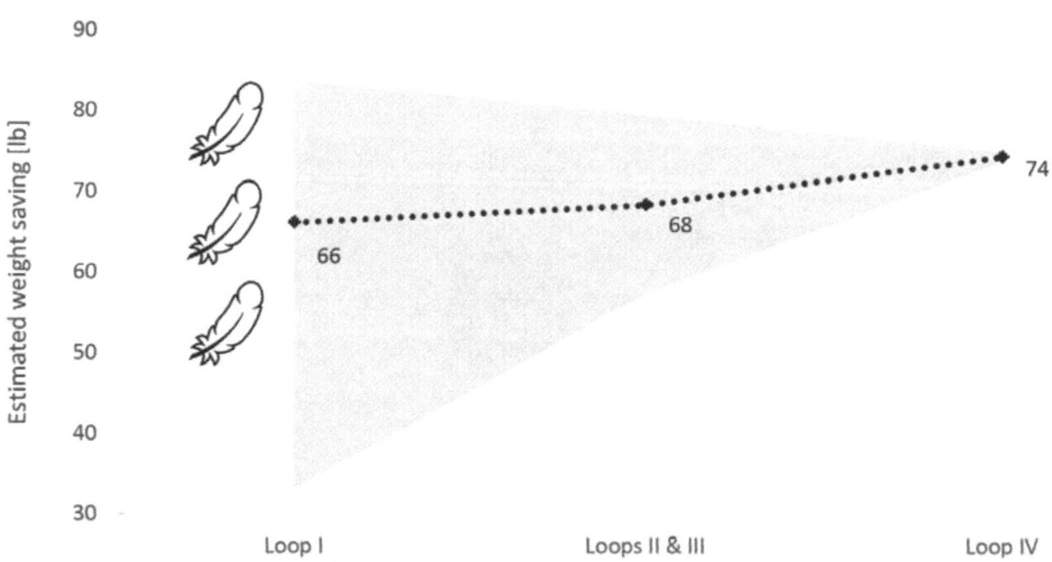

Figure 23: Weight funnel of the new battery concept [lb]

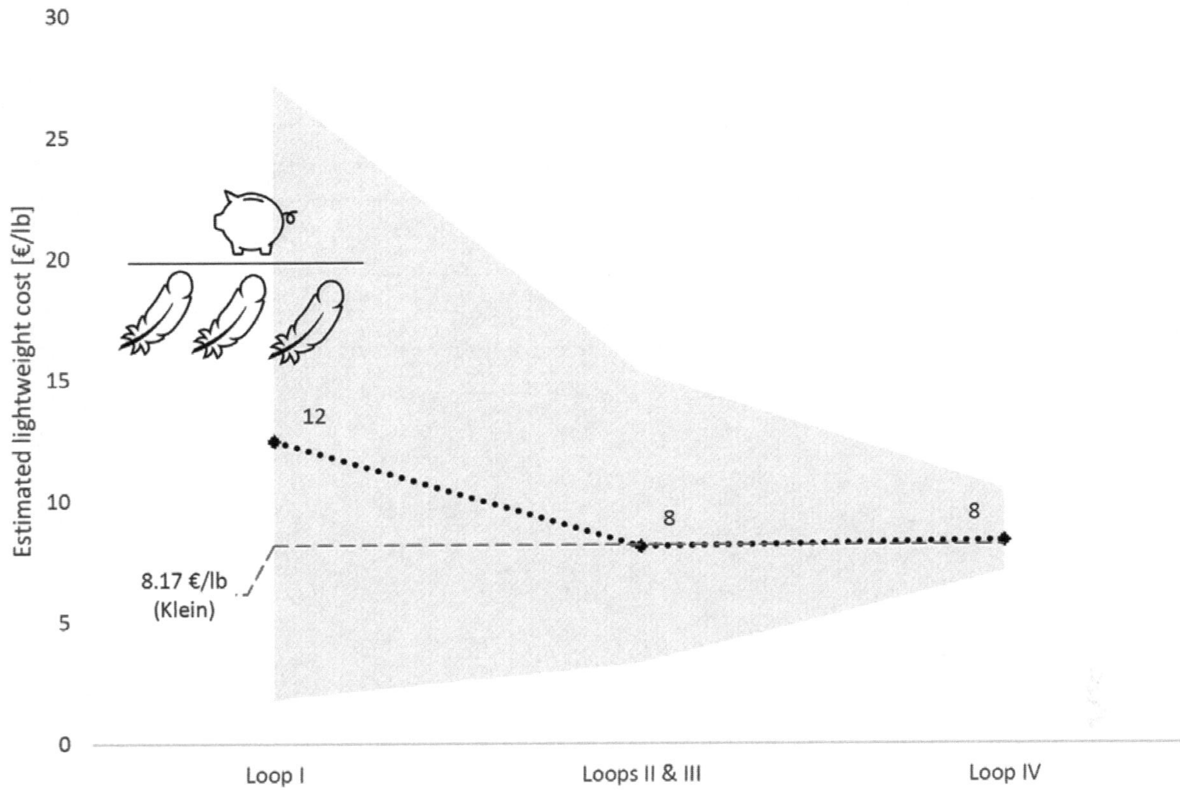

Figure 24: Lightweight cost funnel for new battery concept [€/lb]

Table 5: Weight Breakdown F15 Hybrid [24]

Parts	Weight [kg]
Body	843.810
Interior	109.547
Seats	104.735
Accessories	11.959
Suspension	337.691
Exhaust	39.219
Engine	345.704
Gasoline Fuel Injection System	7.091
Engine Block	31.155
Engine mounts	14.324
Cylinder Head	31.247
Crankshaft system	33.057
Front Engine System	4.912
Lubrication System	14.956
Cooling System	4.131
Turbo/Supercharged System	10.237
Hybrid/Electric System	192.107
Battery System	116.231
Battery	104.142
Single/Left Support	0.551
Front Support	8.431
Rear support	2.642
Rear cover protection	0.055
Central Battery Protection Cover	0.364
Hybrid Battery Drainage Plug	0.046
Power Inverter Module/Converter Module	7.178
HV charger system	18.111
Cooling system	6.930
Rotor System	27.120
Cables System	11.065
Plug-In System	4.048
HV cable bracket system	0.732
Cable Guide	0.274
HV cable cover protection system	0.418
Style Cover on Engine	1.879
Sound Insulation System	0.608
Transmission System	168.270
Brakes Mechanism	83.900
Air System	10.247
Clutch System	14.488
Heating System	28.002
Air Conditioning System	11.666
Steering System	27.268
Cooling System - Water	22.324
Safety System	18.853

Parts	**Weight [kg]**
Fuel System	32.701
Pedals System	2.647
Fluids	99.821
Cooling System - Water	22.324
Safety System	18.853
Fuel System	32.701
Pedals System	2.647
Fluids	99.821
Electrical	98.942
Battery	46.557
Battery	25.971
Battery Fastener	0.431
Battery second fastener (lower)	0.068
Pin +	0.483
Pin -	0.258
Fuses box on battery	0.086
Fuses box cover	0.055
Fuses box support	0.153
Auxiliary battery system	17.547
Pin + Auxiliary Starter	0.188
Pin + Auxiliary Starter Support	0.039
Auxiliary Pin + Lid	0.008
Pin - Auxiliary Starter	0.022
Battery Draining Pipe	0.025
Power Distribution Module	0.271
Power Distribution Module Support	0.108
Front Cover Power Distribution Module cover	0.079
2nd Power Distribution Module	0.090
Support Module 2nd Power Distribution Module	0.043
Rear Power Distribution Module	0.136
Support Module 3rd Power Distribution Module	0.266
4rd Power Distribution Module	0.230
Wiring	36.224
ECU Electronic Control Unit	5.350
Power cables	0.527
Mass cable	0.375
Fuses Box/Relay System	1.410
Front Power System	6.178
Central Locking System	0.085
Wiring support system	2.005
Wiring Harness Guide	0.211
Bus Ring Break Plug (Optical Shortcut Connector for Diagnostic)	0.020
	2.411.794

References

[1] Society of Allied Weight Engineers, Inc. (2018) T E C H N I C A L O V E R V I E W Document No. SAWE TO-1:2018: OVERVIEW OF MASS PROPERTIES ENGINERING FOR VEHICLE SYSTEMS, -, Los Angeles

[2] Project Management Institute (2017) A guide to the project management body of knowledge: (PMBOK guide), Sixth edition. PMI global standard. Project Management Institute, Newtown Square, Pennsylvania, USA

[3] Society of Allied Weight Engineers, Inc. (2009) RECOMMENDED PRACTICE NUMBER RP 100: Vendor Weight Control for the Aircraft Industry, 1.0th edn. RECOMMENDED PRACTICE, Los Angeles

[4] Dahm H-P (2007) Computer aided weight and cost management in vehicle and aircraft industry: SAWE Paper No. 3424, Spain, Madrid

[5] Society of Allied Weight Engineers, Inc. (2016) RECOMMENDED PRACTICE NUMBER A-3: Mass Properties Control for Space Systems, A, Los Angeles

[6] (2019) The standard for risk management in portfolios, programs, and projects. PMI global standard. Project Management Institute, Newtown Square Pennsylvania

[7] United States Government Accountability Office (2009) GAO Cost Estimating and Assessment Guide: Best Practices for Developing and Managing Capital Program Costs

[8] (2019) Practice standard for project estimating, Second edition. Global standard. Project Management Institute Inc, Newtown Square Pennsylvania

[9] (2015) Standard: Mass Properties Control for Space Systems (ANSI/AIAA S-120A-2015). American Institute of Aeronautics and Astronautics, Inc, Washington, DC

[10] International Organization for Standardization (2007) Space systems — Mass properties control: Systèmes spatiaux — Contrôle des propriétés de masse(ISO 22010:2007(E))

[11] William Boze, Patrick Hester (2009) QUANTIFYING UNCERTAINTY AND RISK IN VEHICLE MASS PROPERTIES THROUGHOUT THE DESIGN DEVELOPMENT PHASE: SAWE Paper No. 3468, Wichita, Kansas

[12] Alonso E, Lee TM, Bjelkengren C et al. (2012) Evaluating the potential for secondary mass savings in vehicle lightweighting. Environ Sci Technol 46:2893–2901

[13] Kalpakjian S, Schmid SR (2017) Manufacturing processes for engineering material, 6a ed. Pearson Education, Hoboken

[14] Porter ME (1998) Competitive advantage: Creating and sustaining superior performance ; with a new introduction. Free Press, New York

[15] Klein B, Gänsicke T (2019) Leichtbau-Konstruktion: Dimensionierung, Strukturen, Werkstoffe und Gestaltung, 11th ed. Springer Vieweg. in Springer Fachmedien Wiesbaden GmbH, Wiesbaden

[16] BMW (2015) Der BMW X5 xDrive40e.: Pressemappe. https://www.press.bmwgroup.com/deutschland/article/detail/T0207284DE/der-bmw-x5-xdrive40e?language=de. Accessed 04 Dec 2019

[17] Wallentowitz H, Reif K (eds) (2006) Handbuch kraftfahrzeugelektronik: Grundlagen, komponenten, systeme, anwendungen, 1. Auflage. ATZ/MTZ-Fachbuch. Vieweg, Wiesbaden [Germany]

[18] Budde-Meiwes H, Drillkens J, Lunz B et al. (2013) A review of current automotive battery technology and future prospects. Proceedings of the Institution of Mechanical Engineers, Part D: Journal of Automobile Engineering 227:761–776

[19] Andreas Ahlswede (2019) Entwicklung der EBIT-Marge der weltweiten Automobilzulieferer in den Jahren von 2007 bis 2019

[20] Florian Wätzold (2020) Konzeptbewertung der Leichtbaupotentiale für Hybridfahrzeuge bei Substitution von herkömmlichen Blei-Starterbatterien durch Lithium-Eisen-Phosphat-Technologie

[21] Florian Wätzold (2021) Kostenkalkulation einer Batterie mit modularem Gehäuse: Die Herausforderung der realistischen Bewertung von Leichtbau-Kosten, VDI-Z BD 163 (2021)

[22] Jose Miguel Lopez Ascaso, Dr. Christian Rosenkranz (2020) Bleisäurebatterien mit fortschrittlicher AGM-Technologie (Absorbent Glass Mat). https://www.varta-automotive.de/de-de/technologie/agm-batterietechnologie. Accessed 27 Dec 2020

[23] Greg Kable (2014) BMW X5 eDrive prototype first drive review: The X5 eDrive will be BMW's first plug-in hybrid. It's quick and, should it be priced to compete with comparable diesel offerings, will represent an intriguing choice. https://www.autocar.co.uk/car-review/bmw/x5/first-drives/bmw-x5-edrive-prototype-first-drive-review. Accessed 04 Dec 2019

[24] A2MAC1 (2019) A2Mac1: Automotive Benchmarking. A2MAC1 EURL

SAWE Paper No. S3774
Category No. 13, 24
Category No. 35

WEIGHT CONTROL FOR FLOATING WIND INSTALLATION

A.P. Crowle *, P.R. Thies

*Email: ac1080@exeter.ac.uk

University of Exeter, College of Engineering, Mathematics and Physical Sciences
Renewable Energy Group, Penryn Campus, Treliever Road, TR10 9FE, UK.

For Presentation at the
80th SAWE International Conference
on Mass Properties Engineering
October 2-6, 2021

Cocoa Beach, Florida

SUMMARY

Floating offshore wind is a growing market within the renewable energy sector. The floating offshore wind turbines give access to deeper water sites, with minimal visual impact from land. The paper includes the weight control requirements for Spars, barges, semi submersibles and Tension Leg Platforms (TLPs) as floating wind platforms.

There are weight control challenges for the various substructure types during the temporary phases of construction and offshore installation. An accurate assessment of the buoyancy of the floating wind turbine for different drafts and trims is required. Allowances need to be included in the weight calculation for temporary buoyancy, sea-fastenings and grillage.

Weight control for installation has an influence on the weather window for the floating substructures during transportation to the offshore site and mooring and electrical connection. The paper will cover weight calculation methods during early design, detailed design, construction, installation, operation and demolition.

The installation process for a floating wind turbine varies with substructure type and this paper will give an overview of the weight control requirements for loadout, ocean transport and mooring connection. The floating offshore wind turbine weight and centre of gravity has a direct bearing on draft, intact stability and motions. As part of the weight control process the centre of gravity and radii of gyration need to be accurately determined for each stage of the installation.

1. Introduction

Floating offshore wind turbines are being developed to produce renewable energy in water depths beyond about 200 and up to 1,000 ft (61 to 305 metres). The options for floating offshore wind turbines are described in section 2. The design requirements for weight control are described in section 3. Weight reports during various the construction phases are described in section 4. Weight measurement during the construction and installation phase is presented in section 5.

2. Structures

2.1 Size and Complexity

The three main types of floating offshore wind turbine are) are shown in figure 1 and are the semi submersible, the Spar and the Tension Leg Platform (TLP) and are certified by classification societies per ref [1], [2], [3] and [4]. Damping barges have similar installation methods to those used for semi submersibles.

Floating Offshore Wind Turbine (FOWT) structures and their associated subsea cables are very capital intensive. The topsides of tower, nacelle-generator and blades are very similar to those used on fixed offshore wind platforms and thus their weight and vertical centre of gravity are readily available. FOWT weight control is carried out in accordance with the international guidelines, ref [16].

Table 1 gives details of the weight constraints per type of Floating Offshore Wind Turbine, which determines the selection of construction and installation methods. During construction of the floating offshore wind turbine weight control is very important to ensure that draft constraints during fit-out and offshore towing are not exceeded.

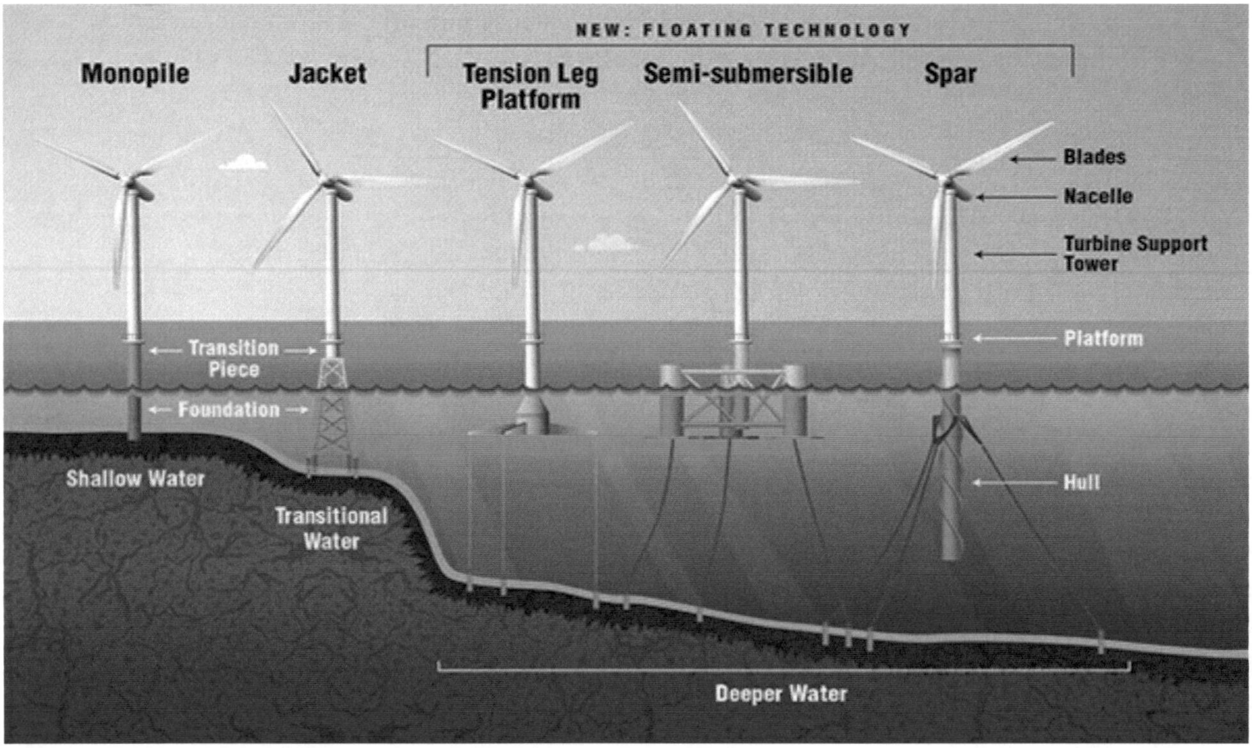

Figure 1 Floating offshore wind turbine types (Ref [5])

Table 1 Weight related constraints for various types of FOWT

TLP	SPAR	BARGE	SEMI SUBMERSIBLE
Very Low intact stability during tow-out, ref [6] and [7] so probably requires temporary buoyancy	Needs solid ballast after upending, ref [8] and [9]. Deep water required for inshore construction	many mooring lines, ref [10]	Heavy construction. Temporary buoyancy in dry dock, ref [11] and [12]

2.2 Floating wind turbine components

Floating offshore wind structures must be moved, floated, and rotated to attitudes differing from those under which they were initially built, thus it is important to have accurate weight data for all the different components. The main components of a floating offshore wind turbine are:

- Substructure is the hull (Spar, TLP, barge or semi-submersible)
- Foundations secure the FOWT to the seabed (driven or drilled or suction pile, drag anchor)
- Mooring lines are chain, wire or synthetic fibre or a combination of these.
- The topsides are the tower, nacelle=generator, hub and blades
- There are inter array cables connecting the FOWT to the subsea export cable

3. Weight control in design

3.1 Inputs

Weight has a direct bearing on draft, intact stability and motions of the floating offshore wind turbine. In addition the centre of gravity and radii of gyration need to be accurately determined.

The input data for the weight report is based on:

- Weight estimates from previous designs
- Material take off from drawings
- Supplier weight quotes
- As the design develops actual weights from suppliers
- Solid ballast in the base

In addition for each weight items the following need to be developed:

- Tolerances on weighed items, e.g. steel plate rolling tolerance of +/- 2%
- Allowances for unknown information
- Contingencies

3.2 Hydrostatics

Hydrostatics of the floating offshore wind turbines are required for different drafts and trims. Due account needs to be taken for variations in water density at the shipyard and fit out quay. The output includes:

- Buoyancy
- Centre of buoyancy
- Metacentric height

3.3 Analysis Steps

- Approve the weight control procedure
- Agree on weight contingencies and reserves
- Agree on weight budget and not-to-exceed weight
- Agree range of CoG, in three directions
- Determine total weight (reported weight + contingencies + loads)
- Confirm that the total weight is less than the not-to-exceed weight
- Confirm that the CoG is within the allowable range
- Issue weight report

3.4 Weight report information

Weight reports are issued on a regular basis, or when there are major changes to the structure. Throughout the project, periodically report

- The status of the overall weight and it's individual components,
- Changes to component weight and CoG from ocurrent report to the previous report,

- Trend of the base weight growth,
- Contingency, allowance reduction,
- CoG location within the CoG envelope,
- Allowable VCG curve
- Radii of gyration
- Buoyancy of the substructure
- Change in weight budget and loads,
- Changes in the weight reserves.
- Flag early warning of any potential over-weight-budget situation

Each weight report, during the design phase, is a snapshot of current knowledge

Weight control will provide continuous reporting of weight and centre of gravity to ensure that:

- Weights and centre of gravity are accurately predicted in detailed engineering
- Lift weights do not exceed the available lift capacity
- Structural design can proceed effectively
- The centre of gravity for each structure does not move outside the envelopes agreed

As the project develops the actual weights are calculated from design drawings and vendor weight data to develop the Net weight. Over the life of the project the net weight replaces the estimated weights. In addition allowances are made from the start of the project, which are gradually reduced as more detailed information becomes available:

- Initial estimated weight and contingencies
- Design development replaces estimated weights with calculated values
- Fabrication may use different materials
- The contingency factors are reduced as the information becomes more accurate
- The nacelle-generator will be weighed before delivery to the fit-out quay
- Weighing the complete substructure before loadout reduces contingency factors.

4.0 WEIGHT CONDITIONS

4.1 Loadout

Loadout, figure 2, weight and CoG is applicable when the item is to be moved horizontally from the shipyard onto the transport vessel. Installation aids and guides, rigging, or other temporary items are included. This weight is used to check transport vessel intact stability and strength.

Figure 2 Loadout of semi-submersible, ref [11], and Spar ref [8]

4.2 Floatout

Floatout, figure 3, is the weight applicable when the FOWT is built in a dry dock. Weight, buoyancy and their associated centres are critical elements in floating out of the dry dock.

Figure 3 Loadout of steel semi-submersible, ref [11], and concrete Spar ref [8]

4.3 Dry Transport

The transport weight, figure 4, of the substructure during the dry tow phase includes any temporary items such as sea-fastenings and grillage weights are to be included. Weight and vertical centre of gravity is critical for assessing the transport vessel intact stability.

Figure 4 Dry transport of semi-submersible, ref [11] and ref[12]

4.4 Floatoff

Floatoff, figure 5, of the substructure hull after dry tow on a heavy transport vessel (HTV). Note that the heavy transport vessel is ballasted to a deep draft and it is imperative that the draft and hence the weight, buoyancy and their centres of the FOWT substructure are acceptable.

Figure 5 Floatoff of semi-submersible, ref [11], and Spar ref [8]

4.5 Fit out

At the fit-out quay, or sheltered inshore waters, the topsides are added to the substructure, figure 6. Installation aids, guides, or other temporary items such as rigging are included in the weight.

Figure 6 Fit out of semi-submersible, ref [11], and Spar ref [8]

4.6 Tow out

Tow-out weight and CoG, figure 7, is important in determining the draft of the structure for the tow to the offshore location phase. It is to include any temporary items.

Figure 7 Wet tow of semi-submersible, ref [13] and damping barge ref[10]

4.7 Operating

The Operating weight, figure 8, applies following hook-up and commissioning, when the complete structure operates under normal conditions; i.e. when storage tanks and pipework contain related fluids and consumables, at designed maximum operating levels. The storm weight is assumed to be the same as the operating weight as the floating offshore wind turbines are designed to be unmanned but to remain stable in all operating conditions.

Estimates of mooring equipment are to be calculated but shown as separate items. Estimates of dynamic mooring cables are to be calculated but shown as separate items. Estimated for variables such as marine growth, ice or snow are not included in the weight report however they form part of the naval architecture and structural analyses. One compartment damage is not included in the weight report but form part of naval architecture analyses.

Figure 8 Semi submersibles operating, Ref [13], Ref [11] and Spar operating Ref [8]

Prior to the demolition of the FOWT estimates need to be made of items added or removed during the operation life.

5.0 Weight items

5.1 Radii of gyration

Radius of gyration of a body about an axis of rotation is defined as the radial distance to a point which would have a moment of inertia the same as the body's actual distribution of mass, if the total mass of the body were concentrated. The radius of gyration of a floating body is important input into any motions calculations for the dry tow, wet tow and operations.

5.2 COG

For each construction phase the transverse COG (TCG) and longitudinal COG (LCG) are to be within the range set by naval architecture and structural calculations.

The allowable vertical centre of gravity (VCG) curve is developed by the naval architect. The maximum allowable VCG curve is a combination of the limitations resulting from the intact stability and the damage stability requireements.

5.3 Weight allowances

Weight contingencies are to be developed, example for example in table 2

Table 2.1 Weight contingencies substructure

Item	Source of Margin	First Estimate	End of Detail design	Weighing
		%	%	%
Sub structure, hull	General tolerance error	5	2	2
	Potential design error	4	0	0
	Fabrication – mill tolerance, welding	3	3	0
	TOTAL MARGIN	12	5	2

Table 2.2 Weight contingencies topsides

Item	Source of Margin	First Estimate	End of Detail design	Weighing
		%	%	%
Topsides Tower, nacelle, blades	General tolerance error	8	2	2
	Potential design error	7	0	0
	Fabrication – mill tolerance, welding	3	3	0
	TOTAL MARGIN	18	5	2

6. Weight measurement during construction and installation

6.1 Design input

Theoretical calculations will typically be performed in the engineering office of the shipyard based on approved for construction. A margin or allowance will be carried to account for the missing details until such time as they are defined.

6.2 Weighing the completed substructure

The Fabricator will produce a weight and CoG estimate based on fabrication drawings on a per-component basis. Weight control is a continuing activity, based on a hybrid of design drawings and fabrication drawing take offs. There is a physical weighing of the substructure prior to loadout.

6.3 Draft survey

A draft survey is required just after arrival at the fit-out quay and just before departure for the offshore site. The following are measured:

- Draft mark on each column to an accuracy of better than +/- 1 inch (25mm)
- Sea water density at the surface and 30ft depth
- Fluid in ballast tanks
- Assessment of solid ballast added at the base

From these measurements the following can be determined using the hydrostatics:

- Weight of structure
- Trim
- Heel
- Longitudinal centre of gravity
- Transverse centre of gravity

6.4 Possible inclining experiment

An inclining test, ref. [2] may be used to determine weight, plan CoG and in particular the vertical centre of gravity of the FOWT at the fit-out quay.

7. Discussion And Conclusions

7.1 Discussion

The weight and COG of the floating offshore wind turbine is direct input into the draft and intact stability at various stages of construction. The weight, COG and radii of gyration are important inputs into the motions during the dry tow, wet tow and operations. Spars have solid and water ballast. Semi submersibles have water ballast.

7.2 Conclusions

Accurate calculations are required for weight, CoG, radii of gyration and buoyancy at the various stages of construction, installation and operation.

8. References

[1] DnV ules For Classification Floating Offshore Wind Turbine Installations, 2020
[2] ABS Guide For Building And Classing Floating Offshore Wind Turbines, 2020
[3] Bureau Veritas Classification of Floating Offshore Wind Turbines 2019
[4] Class NK Guidelines for offshore floating wind turbine structures, 2012
[5] Bureau of safety and environmental enforcement, 2020
[6] 'www.bw.com photos, 2020
[7] 'www.sbm.com photos, 2020
[8] 'www.equinor.com photos, 2020
[9] 'www.stiesdal.com photos, 2020
[10] 'www.bw-ideol.com photos, 2020
[11] 'www.principlepowerinc.com photos, 2020
[13] 'www.wison.com photos, 2021
[15] 'www.umaine.edu, 2013
[16] ISO 19901-5:2016 Weight control during engineering and construction

9. Acknowledgements

Alan Crowle, a research student into the 'Installation of floating offshore wind turbines', thanks his colleagues at the University of Exeter for their assistance in preparing this paper. Professor PR Thies, renewables department at the University of Exeter, would like to acknowledge the support through the EPSRC Supergen ORE Hub [EP/S000747/1].

10. Terminology

CoG: Centre of gravity
FOWT: Floating offshore wind turbine
LCG: Longitudinal centre of gravity
SPMT: Self propelled modular transporter
TCG: Transverse centre of gravity
VCG: Vertical centre of gravity

AUTHOR INDEX

Society of Allied Weight Engineers, Inc.
Ronald L. Fox
Executive Director
5734 E Lucia Walk
Long Beach, CA 90803-4015
USA

ISBN 978-1-7138-4499-0

CURRAN ASSOCIATES INC.
proceedings.com

European Technical Coatings Congress (ETCC 2022)

Coatings Inspirations

Krakow, Poland
12-14 July 2022